BENDING THE RULES

BENDING THE RULES

MORALITY IN THE MODERN WORLD

From Relationships to Politics and War

ROBERT A. HINDE

With contributions from
SIR JOSEPH ROTBLAT

OXFORD

UNIVERSITY PRESS

OXFORD
UNIVERSITY PRESS

Great Clarendon Street, Oxford OX2 6DP

Oxford University Press is a department of the University of Oxford.
It furthers the University's objective of excellence in research, scholarship,
and education by publishing worldwide in

Oxford New York

Auckland Cape Town Dar es Salaam Hong Kong Karachi
Kuala Lumpur Madrid Melbourne Mexico City Nairobi
New Delhi Shanghai Taipei Toronto

With offices in

Argentina Austria Brazil Chile Czech Republic France Greece
Guatemala Hungary Italy Japan Poland Portugal Singapore
South Korea Switzerland Thailand Turkey Ukraine Vietnam

Oxford is a registered trademark of Oxford University Press
in the UK and in certain other countries

Published in the United States
by Oxford University Press Inc., New York

British Library Cataloguing in Publication Data
Data available

Library of Congress Cataloging in Publication Data
Data available

Typeset by SPI Publisher Services, Pondicherry, India
Printed in Great Britain
on acid-free paper by
Biddles Ltd., King's Lynn, Norfolk

ISBN 978-0-19-921897-4

1 3 5 7 9 10 8 6 4 2

ACKNOWLEDGEMENTS

In 2003 a book by the late Sir Joseph Rotblat and myself on eliminating war was published. Later that year, and just after his ninety-fifth birthday, Jo had a mild stroke. While in hospital he came to feel that he was never going to be able to engage in constructive work again. So I suggested that we should write another book together. It was to start from two papers he had written for a meeting of Nobel Peace Laureates, which I had read for him because he was unable to travel to Rome. One was on ethics in science and the other on ethics in politics. Those two papers form the bases of Chapters 5 and 7 of this book. When Jo recovered from his stroke he became deeply immersed again in his many concerns—the abolition of nuclear weapons, the elimination of war, the question of prisoners of conscience, the death penalty, and many others. Consequently, he had no time for writing about ethics in other fields. He therefore felt he should not remain as an author to this volume, but after considerable discussion he eventually conceded to the present form. Writing has led me to a rather different view of morality from that with which I started, but it is my sincere hope that Jo would have approved of the final draft.

In writing these chapters I have been indebted to many colleagues for comments and discussion, especially to Robert Evans, Adam Fisher, Christine Gray, John Finney, Jack Harris, Jane Heal, Jonathan Hinde, Howard Hughes, Christel Lane, Lord Mustill, Benjamin Parker, Michael Pollitt, Joelien Pretorius, Pete Richerson, Malcolm Schofield, Joan Stevenson-Hinde, Sylvana Tomaselli and Eden Yin. None of them bear any responsibility for shortcomings in the final version. I would like also to express my special gratitude to my editor, Latha Menon.

CONTENTS

Introduction

Do-unto-others-as-you-would-have-them-do-unto-you. Some variant of this 'Golden Rule' is basic to our society. Yet in the commercial world people do the best they can for themselves, in war soldiers kill, politicians distort the truth to please the electorate, and barristers defend individuals whom they believe to be guilty or prosecute ones they believe to be innocent. Seen from the standpoint of everyday morality, they are breaking the Golden Rule, yet not only does the businessman, soldier, politician, or barrister believe him- or herself to be doing the right thing but, to different degrees, their behaviour is accepted by others. Are the moral rules we live by, and that seem to be absolute, more flexible than seems at first sight? If so, is the flexibility imposed by our society or is it an inevitable consequence of the human condition?

These are important questions, for if the pillars of society can bend the rules, so can every Tom, Dick, and Harry, and perhaps this is at the base of the malaise from which society often seems to suffer. In this book I argue that the complexity of society, our propensity to divide our world into in-group and out-groups, and our human desire to see ourselves in the right can indeed lead to our bending the rules. In some cases the revised morality is assumed only by the individual concerned, in some it is acknowledged only by an in-group, and in yet others it is publicly accepted.

It is necessary first to understand the origins and nature of moral codes and values. I will argue that the role of morality[1] is to maintain a balance between two categories of human behaviour, prosociality (roughly, actions that are beneficial to others) and selfish assertiveness (behaviour that furthers one's own interests, regardless of others). I shall say more about these categories later, but each involves, at a finer level of analysis, innumerable types of behaviour.

If that were all there were to it, we could understand the bad behaviour of others by simply saying that selfish assertiveness was dominating their behaviour. But human behaviour is multilayered, so that behaviour we see as right in one context may appear differently in another. Indeed, a person may think she[2] is behaving well when we think she is not. Is she merely deceiving herself, or is she judging her behaviour against a morality that differs from our own?

That is the question I set out to answer. It has led me to a somewhat different view of morality from that which has been purveyed down the centuries—different not so much in what is seen as good or bad, but in the nature of morality. Although this approach leads to the view that we sometimes adjust the rules to make ourselves feel good, it is certainly not a relativist one in the sense that anything goes, just make up your morality and live by it, for two reasons. First, the precepts by which people see themselves to be living must be perceived as compatible with basic principles, including the Golden Rule, that make group life possible. And second, moral codes are essentially social matters: precepts are meaningless unless accepted by the group to which one belongs.

Many will say that there are too many books about morality already: the bases of morality pose problems that have occupied some of the greatest minds for thousands of years, from Aristotle through Bentham, Hume, and Moore to Rawls, Bernard Williams, and other contemporary philosophers. But even the philosophers themselves are not entirely happy with the results of the academic infighting to which moral philosophy has given rise,[3] and perhaps there is a case for a new look. Can rational discourse be grounded more firmly in what really happens in the world?

This book is based on an ethological concern with what people actually do and value. It depends in part on earlier evolutionary approaches to morality,[4] and uses more recent work that has pointed the way towards the solution of a problem that has long posed a difficulty for evolutionary theories, namely how could altruism and prosocial behaviour have evolved in nature seen as 'red in tooth and claw'?[5] But a biological approach is insufficient to provide a full understanding of moral codes, and I have drawn also on data from psychology, anthropology, and comparative studies of religion. In doing so, I focus on the rules that people feel they ought to follow and the values they feel they should hold.[6] Thus I avoid the implication that morality necessarily concerns what is right and just in any absolute sense. Though using an approach based in both biological and cultural evolution, I am certainly not implying that what is natural is right, or that what is right is natural.

Some will contest my implied claim that ethics is not a matter solely for theologians and philosophers. Science, they will say, is concerned with what is, not with how things ought to be: values are held to be outside its realm. By the approach presented here, that view is out of date. Indeed, I shall attempt to show that science, and especially the behavioural sciences, has a great deal to offer towards understanding moral values and behaviour. It is not good enough to say 'I just know what is right', or to ascribe our knowledge of good and evil to 'moral intuition'. We have to ask what 'moral intuition' means.

I argue that our moral rules come from our human nature in interaction over time with our social environment, and are stored by individuals so as to become virtually part of their natures. Morality is neither a direct product of our evolutionary past, nor is it simply something we learn, but a product of mutual influences between them. As part of our make-up, its bases are constant across cultures and across social environments. Because they are partly acquired, moral codes can differ between cultures and even, to a limited extent, between situations. They are a product of natural and cultural selection because they have helped to maintain a balance between our prosociality and our selfish assertiveness, thereby making social life possible.

The route I have followed has not been the traditional philosophical one of reflection on how we ought to behave, but focuses on how people behave and what people value in the real world. Using insights from the behavioural sciences, I ask four interrelated questions. How do individuals acquire their moral outlook? What causes an individual to adopt (or reject) a moral course of action? Over recent, historical, and prehistorical time, how did moral rules and values evolve? And how do they function, in the sense of how is it helpful for this to be seen as good and that as bad?[7]

In the real world, moral conflicts abound. Consciously or unconsciously, we encounter them whatever we are doing. Examination of a number of contexts in the modern world exposes how people can behave questionably but nevertheless perceive themselves as doing the right thing. It is my hope that this approach is a step towards providing a deeper insight into the Socratic question of how we should live.

This book is organized as follows. The first chapter outlines an approach to the bases of morality that stem from the human sciences and discusses how morality develops in the individual. Chapter 2 describes the role of long-term human evolution in morality. Chapter 3 is concerned with the relatively short-term changes in morality brought about by cultural evolution. Many of the data here come from studies of the evolution of legal systems, so the relations between morality and law are discussed. Chapter 4 is concerned with close personal relationships, the context in which proto-morality first evolved. Even in close relationships, disagreements and conflicts often arise. This is followed by five chapters emphasizing the ethical conflicts that arise in different spheres of modern life, as seen by an outsider. I emphasize 'as seen by an outsider' not solely because my outsider's view is inadequate, which is certainly the case, but because the insider may see things differently. That, indeed, is the point. In each context of our lives the issues are different, and many of the problems that arise stem from differences between the guidelines required for tension-free personal relationships and those constructed in the several spheres of modern life. A concluding chapter focuses on the anomaly that arises from the necessity that moral precepts should be seen as more or less absolute, and yet the tension induced by the complexities of life can be reduced because people can think they are behaving well when they are not. Finally, I reflect on the relevance of this approach to morality in the world today.

A brief Appendix summarizes the principal similarities to and differences from the approaches of moral philosophers.

1

Ethical Principles and Precepts

The more fortunate among us live peaceful and productive lives. Yet every day murders, burglaries, and crimes of many sorts are reported in the newspapers. We have become used to corporate scandals in the business world. We know that, while some individuals have more wealth than they know what to do with, millions of others are trying to exist below the poverty line. Wars, bringing death, bereavement, misery, and devastation are going on in other parts of the world.

There can be no simple answer to the ills of the world. But this book focuses on one issue that I believe to be important and that has hitherto been insufficiently stressed. To be specific, behaviour can rather easily get unhitched from the moral code which we assume guides our lives even when we think we are behaving properly. This can occur because the demands of the various contexts in which people live can affect not only their behaviour but also the criteria by which they guide their behaviour. One can convince oneself that one is behaving properly when others think differently.[1]

We must first ask where our ideas about what we should and should not do come from. Much of the time we do not have to think consciously what we 'ought' to do, we do it 'intuitively', though that does not necessarily mean that our intuition is independent of our experience. In other cases we think long and hard, weighing up the consequences of doing this or that, conscious of the rules that should be guiding our behaviour. Sometimes our considerations lead to a rationalization of what we had concluded intuitively, while at other times they involve genuine consideration of the pros and cons. We may, or may not, take the course that we perceive to be the more 'right'. We may also deceive ourselves about which is right: that is a matter to which we shall return.

But how do we know which is the more 'right'? Many ascribe it to 'moral intuition', saying that one just knows what is right. Our intuitions or reflections about actions often refer, directly or indirectly, to their consequences, but I shall seek for their roots not in some abstract quality of those consequences[2] but in the evolutionary and developmental factors that lead us to our moral rules. Before we can discuss the difficulties that people often face in making moral decisions, we must seek answers to these developmental and evolutionary questions:[3] it may be that goodness is what we see as good, not because there is some abstract quality of goodness, but simply because we are what we are.

At first sight, there seem to be only three possible answers to this question of how we know what is right. Some will say that God tells them what is right. Or, that God gave basic moral precepts to Moses, inscribed on blocks of stone, and they have served to guide us ever since. We can neither prove nor disprove that this is the case. Certainly such beliefs provide guidance and comfort to some. But, as MacIntyre[4] has pointed out, it is difficult to specify how theism can be a *source* of ethics, as Christian ethics was largely drawn from elsewhere. In any case, it is difficult for many twentieth-century minds to believe in a transcendental being who is everywhere at once and knows everything: it is incompatible with what we know about the world we live in.[5] So in this book I pursue another tack.

7

Others believe that our knowledge of what is right, of what should be regarded as good and what as evil, is provided by 'culture'.[6] After all, the Ten Commandments have been handed down over thousands of years, and serve as guidelines for much of what we do or do not do. In the West, the churches deserve credit for having been the main purveyors of our morality. Although not its original source, they have preserved it although sometimes subtly changing its emphasis. For instance, Luther's view of Christianity shifted its central emphasis away from the context of pre-Reformation community to the internal life of the individual. Previously salvation had depended on being one of the community of the elect—hence the power of excommunication. Luther saw salvation as depending on individual faith.

But the suggestion that our morals come from culture is hardly satisfactory, because it raises the further question, 'Where does culture come from?' Some elementary forms of culture are known in non-human species, and have been well studied in chimpanzees,[7] but culture as we know it in all its complexity is limited to human beings.

This brings us to the third possibility, that our moral intuitions are products of human nature. As developed here, this is a relatively new approach, as the advances in the behavioural sciences on which it is based have become available only in recent decades.

It is necessary to digress briefly to clarify how I am using the concept of 'human nature', as it has fuzzy edges. By 'human nature' in the strict sense (as contrasted to the loose sense discussed below) I refer to pan-cultural psychological potentials or characteristics possessed (albeit to varying degrees) by virtually all humans or by all members of an age/sex class. Babies brought up in every culture become able to walk, talk, eat, drink, and sleep, and to distinguish good from bad behaviour. (That what they call good may not be what you or I would call good is another matter to which I shall return.) I am making no implication as to the relative roles of our genetic make-up or the environment in the development of these characteristics and potentials. Potentials depend on experience for their realization, but here we are concerned with some that are realized at least to some degree in all environments in which humans live. For instance virtually all adults walk with a similar gait, are startled by loud noises, and eat when hungry. This says nothing about the relative roles of genes and experience in the development of these traits: walking may depend on the infant living in an environment where she is able to pull herself into a standing position, but (virtually) all babies enjoy such an environment.

It is becoming apparent that there is a very large number of these human universals,[8] but for the moment it is necessary to focus on only two: the propensity to do the best for oneself, which I shall call 'selfish assertiveness', and the propensity to please others, to be cooperative, kind, loving, and so on, which I call 'prosociality'. 'Selfish assertiveness' and 'prosociality' are mere labels for global categories, each embracing many types of behaviour that become differentiated in the course of development. In general, selfish assertiveness includes actions that advance the interests of the actor, often but not always to the detriment of others, while prosociality includes behaviours that foster the well-being of others, usually at some immediate cost to the actor.

I shall say more about these categories of behaviour later in this chapter: for the moment I want to emphasize the importance of the human tendency to be prosocial. Partly because of the media reports of murders, muggings, rapes, and so on, it is easy to get the impression that humans are inherently evil. Perhaps the tendency to think that way has been helped by the (now mostly disregarded) Christian doctrine of original sin, which was part of our culture for so long, and by child-rearing practices derived from it. What the papers do not report are the innumerable acts of kindness and cooperation that most of us receive every day.

We have a tendency to be prosocial as well as to be selfishly assertive; if it were not so, social life would be impossible.

Related to prosociality is the so-called Golden Rule of Do-as-you-would-be-done-by. This has a number of alternative formulations: one is Do-to-others-as-you-think-they-would-like-you-to-do-to-them. In one form or another, it is central for probably every society: indeed it is difficult to believe that a society could exist without some such principle. To avoid misunderstanding, I am certainly not saying that the Golden Rule is directly determined by the human genome: it may have been reinvented by every human group that persists, possibly because humans have always lived in groups and group-living would not be possible without such a rule. I call such pan-cultural moral rules *principles*.

But the view that 'moral intuitions' are just part of human nature cannot be wholly satisfactory because both individuals and cultures differ in their morality, and because we know that experience enters into the development of moral decision-making. In any case it is simply not good enough to say that we all know that X is right and that Y is wrong, and build a complicated argument on that basis. If morality is really to be understood we need to know why we all think like that. Why do we not all think that Y is good and X bad? How do individuals develop a sense that X is good and Y bad? Why does whether X is good sometimes depend on the person to whom the action X is directed? And how is it that all cultures may not agree that X is good?

In this chapter I sketch an answer to these questions by considering a slightly more complicated possibility, which still acknowledges the important role of human nature. It requires a distinction between human nature in terms of potentials common to all individuals in every human culture on the one hand (the strict sense as discussed above), and human nature as displayed by the potentials and characteristics common to most humans in a particular culture and circumstances on the other (the loose sense). Behavioural propensities common to all humans may produce behaviour, attitudes, values, and so on that differ between cultures because of differences in the cultural (or physical) environments. It is part of the nature (in the loose sense) of a pygmy to relax in a squatting position, but not in the nature (loose sense) of a European. Similarly with moral rules: an orthodox Jew, but not a Christian, would be horrified if forced to eat pork. But such differences do not imply differences in nature in the strict sense.

Thus in addition to the moral *principles* found in all cultures, we have rules and values for guiding behaviour that are more or less limited to one or a few cultures: the latter I refer to as *precepts*.

Ultimately, these precepts differ between cultures because of differences in circumstances.[9] Human groups have lived in environments ranging from tropical forests to the Polar wastes, and have depended on gathering, fishing, hunting, and agriculture, not to mention all the complexities of industrial societies. What matters in one environment may be irrelevant in another. Human cultures have therefore come to differ in many ways, each having its own customs, myths, precepts, values, and conventions. The rules in small egalitarian groups of hunter-gatherers are very different from those in a modern urban metropolis. Revenge, for instance, is a much more potent precept in some cultures than in our own. Some people see polygamy as acceptable and indeed normal, most Westerners do not. These differences must have emerged over historical and perhaps prehistorical time from the transactions between environments, cultures, and the behaviour of people in each society.

Such differences are sometimes glaringly obvious—for instance differences in the foods that may be eaten, or the relative values placed on respect for individuals compared with the well-being of the community. Each human group has elaborated rules or precepts which, though based mainly on basic principles shared by all cultures, such as *Do-as-you-would-be-done-by*, can differ between groups. The basic principles are ubiquitous and absolute, the precepts have some degree of cultural specificity. For example, many of the Ten Commandments concern prosociality, and are accepted as basic in most Western societies. Injunctions such as 'Thou shalt not kill' and 'Thou shalt not steal' are compatible with the Golden Rule and, as applied to fellow group members, are probably ubiquitous. However '…thou shalt not covet thy neighbour's wife…' would be meaningless in societies where the institution of marriage as we know it does not exist. Amongst the Aché of Amazonia, a woman may have intercourse with many partners, thereby acquiring their subsequent help in rearing the child: this practice is apparently sustained by a belief that the foetus requires repeated doses of semen.[10] Again the Moso, in south-west China, until recently did not officially recognize biological paternity: most brothers and sisters remained in their natal 'House', the men taking a paternal role with their sisters' children and sexual partners meeting at night outside the House.[11]

We know rather little about how such cultural differences arose, but in the long term it is likely that adaptation to local social and environmental conditions, decisions by or examples of leaders, and chance all played a role. In the shorter term the main (but not the only) engine of cultural evolution has been a two-way interaction between what people do and what they are supposed to do. Circumstances influence the ways in which people think and behave, and how they behave affects the culture, which is a special part of the circumstances in question, and this in turn affects how people behave, and so on.[12] Put another way, the moralities that are part of each culture are affected by two-way interactions over time between what people do and what they are supposed to do. We are affected by the moral climate in which we live, though we may either be guided by it or rebel against it, and how we behave affects the moral values held in the community. Culture affects people and people affect culture.[13]

To give a recent example, before World War II divorce was regarded as disreputable in the UK, and divorcees were often stigmatized. After the war, for a variety of reasons, divorce became more frequent. As it became more frequent, it became less disreputable. As it became less disreputable, it became more frequent. While this omits other factors, the point is that how people behaved affected the cultural values, and cultural values affected how people behaved. At the present time we are seeing a similar change in the acceptability and frequency of premarital cohabitation, of four-letter words and nudity in the media, and in countless other matters. Sometimes such changes can be to the detriment of humankind, especially if they are instigated by a temporary change in circumstances, but our cognitive abilities often edge them in directions that are beneficial to ourselves or our group: people do not always follow blindly what others are doing, but select changes in cultural patterns that favour their own needs and desires (pp. 22–3). Anthropological evidence suggests that some (though not all) of the ways in which morality is used to guide individual decisions differ between cultures in ways that appear to be adaptive.[14]

Thus moral precepts and values are not set in stone but rather maintained by a dynamic between processes leading to their creation, their maintenance, and their gradual disappearance. At the same time, to be effective they must be *perceived* as absolute. I shall return to this anomaly in later chapters.

Today, most people live in complex societies, each of which contains many overlapping groups. Each group may construct variations on the societal moral code. Dealings in the business world, for instance, may be seen as requiring precepts differing from those in the family. This can lead to people ordering their business lives by a code that differs to a limited degree from that recognized by other members of the society. They may see their code as moral, and provide reasons to themselves as to why they must follow it, while others accept it at most half-heartedly (Chapter 8). In other words, some of the things that we dislike in the world cannot be explained simply by saying that 'they' are behaving badly: they may believe that they are behaving correctly by rules that differ from those by which we judge them. In general, we judge others against a moral code that has been developed to smooth relationships between individuals living in small groups, but there are circumstances in which people have adjusted the rules, usually but not always to suit their own interests, according to how they see the context. One wonders how far this is an inevitable consequence of the complexity of society.

Later chapters contain examples of such effects: the ethical precepts accepted in different spheres of modern life differ somewhat according to the demands placed on individuals. Most of the cultural differences between societies that are only too obvious in the world today are the result of changes over a much longer period.

Three issues in the preceding discussion require further emphasis. First, morality is essentially a social matter: it would be meaningless for an isolated individual to work out rules that were more than purely pragmatic for his or her own welfare. The social nature of morality means that precepts have been elaborated that may differ between societies, though only to an extent constrained by basic principles such as Do-as-you-would-be-done-by and the duty of parents to care for their children. They may even differ to a more limited extent between groups and individuals within a society. Later chapters in this book show how ethical precepts differ in different spheres of adult life within our own society.

Second, I have argued that people are potentially good as well as potentially bad, prosocial as well as antisocial. To say that we have two propensities—one for prosociality and one for selfish assertiveness—is no more than an heuristically useful approximation and hides much complexity. But, as T. H. Huxley[15] pointed out, the evolution and development of selfish assertiveness is comparable to that of prosociality, though the relevant experiences differ and they lead to different behaviour. These two propensities are used in everyday speech as guides to behaviour in probably every culture, prosociality as 'Do-as-you-would-be-done-by' and selfish assertiveness as 'Do the best for yourself', 'Be yourself, man', or 'Realize your own potentials'.

'Prosociality' and 'Selfish assertiveness' are global terms: as the individual develops they embrace more and more types of behaviour. Both depend on cognitive and emotional development involving faculties important also in other contexts. We still know too little about the early stages of human development to know with certainty which aspects of human psychology would develop in virtually any environment and which are crucially dependent on the particularities of experience. The number of basic propensities that are largely independent of experience is much larger than these two: we shall meet others in later chapters. Later I shall mention the duty of parents to care for young children as an obvious case. Thus the answer to the long-standing controversy as to whether moral rules are to be regarded as absolute or situation-dependent must be both: the basic principle of the Golden Rule is to be seen as absolute in the sense of being pan-cultural, while the precepts are in some degree culture-specific.

As a third point, it is not necessarily what people *do* that we see as good or bad: we make our judgements on what *we perceive* them to be doing. Actions seen as morally correct by the actor, may be judged quite otherwise by others: Palestine's freedom fighters are seen as terrorists by the Israelis, and Israeli shelling of the Gaza strip is seen as terrorism by the Palestinians. Help given to another with the best of intentions may be resented if it diminishes his self-respect. Context is also important, so that donations to a social club may be seen as praiseworthy, donations to a terrorist organization as antisocial.

It may be as well to emphasize yet again that, although the media and the spectacles we wear may suggest otherwise, tendencies to be prosocial are part of our nature. Displaying kindness and consideration to members of your own group is basic to the moral code of probably all societies. At a meeting in Chicago in 1993, at which all the world's major churches and many minor churches were represented, some variant of the 'Golden Rule' was accepted as basic to moral codes.[16] Of course, as over so many issues, one cannot be certain that the generalization is true of *all* societies as the data do not exist, but it is difficult to imagine how a society in which it was not accepted could survive.

Principles that are in essence the same as the Golden Rule have been intrinsic to much religious and philosophic discourse. For instance Jesus insisted 'Love your neighbour as yourself' and Hobbes's Laws of nature include '...be contented with so much liberty against other men, as he would allow other men against himself'. Kant's 'categorical imperative' —'Act only on that maxim by which you can at the same time will that it should become a universal law'[17] is saying act to others as you would they should act to you. To use one of Kant's examples, I must keep a promise because I would like to rely on promises that others make to me. And the Rawls requirement for an ideal social system is such that it should be acceptable to all its members.

However there are difficulties with the way the Golden Rule is stated.[18] First, it can be stated positively ('Do to others as you would have them do to you') or negatively ('Do not do to others what you would not have done to yourself'). In addition, if you are concerned with someone with a value system different from your own, it might be better stated as 'Do to others as you think they would like you to do to them'. But in that case you may be doing something that you think is wrong. For instance, if you believe that killing is wrong, how should you answer an incurable elderly relative who asks for euthanasia? Perhaps the best formulation is therefore 'Love your neighbour as yourself', though ideally 'neighbour' must be taken to mean all humanity.

Most moral precepts are particular instances of, or are compatible with, the Golden Rule. However, as we have seen, precepts may differ between societies while still compatible with the Golden Rule. Thus one answer to the Platonic question 'What is it about an action that permits us to call it just?' is neither divine guidance nor moral intuition, but 'It conforms to the Golden Rule', which itself makes group life possible.

MORAL DEVELOPMENT

This account of moral rules which are perceived as absolute raises the question of how individuals acquire them. The account that follows omits many of the insights that have been gained by developmental psychologists and is intended only as a background to later chapters.

A growing body of evidence shows that development proceeds more or less independently in a number of distinct domains, such as language, the physical world, the social world, and so on.

Within each domain, children seem to know more than they can have learned: for instance, in the physical domain very young children act as if they 'know' that an unsupported object will fall, and that one solid object cannot pass through another. Cognitive development depends on such basic predispositions, likely to be pan-cultural, that influence the way that children acquire knowledge as they interact with aspects of the animate and inanimate environment. We know little about how far these predispositions depend on previous experience, but if they do, the experience must be common to all babies.

The development of the ability to act morally probably depends on a number of such predispositions, many of which are important also in other contexts: there is some evidence that a predisposition to respond with disgust to certain situations may play a part. However, in general it is clear that knowledge of which actions are good and which are bad is crucially influenced by experience in development, and the development of a human infant requires care of the kind normally provided by a mother. For that reason, accounts of stages in human development that imply that cognitive and moral faculties simply unfold as the child grows are misleading: relationships with others are crucial.

We think of moral development as the acquisition of 'good' behaviour and 'good' values, but comparable processes can and do also lead to 'bad' behaviour and values. All parents agree that babies can be 'bad'. They make messes, spit out their food, cry, and disturb their parents. But these things are not evil, they are what babies do as they develop in the world. The penetrating nature of a baby's cry impels her parent to action and to provide what the baby needs. We can, of course, say that such behaviour is symptomatic of the infant's assertiveness, but infants must express their needs and satisfy their physical urges. They have been adapted in evolution to do so, and indeed would not survive if they did not express themselves. In the early months they can have little awareness that their behaviour may be contrary to their parent's wishes.

It is only with considerable cognitive development that we can begin to talk about *wilful* selfish assertiveness.[19] It is natural for a child to make demands that the parent may not want to meet, but the infant has to be quite sophisticated before one can have evidence that he or she knows that its demands are contrary to its parents' wishes. Nevertheless, in time selfish assertiveness does become a reality. It is important for children sometimes to be able to assert their autonomy, just as every adult is only too aware that there are times when one feels one must have one's own way, no matter what the consequences are for others or, for that matter, for oneself.

That is one side of the picture. The other is that even very young children, less than a year old, show indications of behaviour that could be considered as morally good, such as sharing, taking turns, helping and cooperating, and obeying their parents.[20] Soon after that age they may start to respond to the distress of others with attempted prosocial interventions.[21] While such behaviour shows that young children are predisposed to develop behaviour that we should call moral, it certainly does not indicate that they have a fully fledged moral capacity. But the evidence suggests that they are predisposed to learn to please and help their parents, so long as doing so does not conflict too much with their own interests. Parental smiling and 'Good girl' are potent reinforcers for a child's behaviour. (Indeed, approval or disapproval by others has a powerful influence on our behaviour throughout life.) We can see evolutionary continuity in this, for many young animals must respond to their parents' alarms, and cling to or follow their mothers when the latter start to move: if they did not, they would be left unprotected from predators. It is also in the interests of human infants to please their parents, because then their parents are more likely to satisfy their needs.

But it requires considerable cognitive development before we can talk of genuinely moral behaviour. Children must first acquire an idea of what they themselves are like as individuals—a 'self-system' (see below).

So babies have potentials to seek after their own goals, to do the best for themselves, and also to please and cooperate with others. Both are part of their heritage. Twin studies suggest that a proportion of the variance in prosociality is accounted for by genetic factors, but as yet we know little about how those genetic factors work. They might, for instance operate by influencing the readiness to respond to parental requests.

Further development depends on many aspects of the babies' cognitive development that are important also in other aspects of their lives. For example, morality would be largely meaningless if children were not able to infer the thoughts and feelings of others as well as becoming aware of their own behaviour, intentions, and feelings, or if they were not able to form expectations, or to feel empathy with another's emotions, or to use language to communicate and to formulate their feelings. Early on children may acquire knowledge about moral precepts without the motivation to accept the rules as binding on them: only later are moral standards internalized. Important also is the ability to take context into account: children must learn that they may turn cups upside down when playing in the sand but not at the dinner table.

These issues will not be discussed here: the important issue for present purposes is the development of a balance between prosociality and selfish assertiveness. A major factor in determining this balance is the nature of the parenting that the child receives. Sensitive parenting builds on the propensity for prosociality to achieve a situation in which prosociality tends to predominate over selfish assertiveness. Simplifying considerably, children who have experienced authoritative and reasoned control coupled with tender loving care, each sensitive to the child's needs, tend to be less selfishly assertive, less aggressive, and more prosocially disposed than children who have received either a harsh or a neglectful upbringing.[22] Insofar as morality is related to religious orientation, it has been shown that differences in caregiver sensitivity and the security of the child's attachment to its parent affect the extent to which the parents' religious orientation is passed on.[23]

Parental influences are not just the result of their praise and admonition, but are purveyed also by their style of behaviour, and by the cultural myths and stories that they make available for their children. The chain of causation can be traced even further back: for example, parents living in very deprived conditions, where life is a constant struggle, may find it difficult to be sensitive caregivers. So also, because of other distractions, do some of the very rich. And some parents may encourage selfish assertiveness as the only way to survive in a harsh world, or in the hope that their children will emulate their own selfish assertiveness and become rich too.

Parents are also influenced by the culture in which they live, that is, by the attitudes, values, conventions, skills and so on prevalent in the society. They usually hand on, or try to hand on, the norms and values they have acquired during their lifetimes, and their behaviour influences in turn the prevailing culture. For instance, in some societies masculine assertiveness and feminine modesty are valued, and parents try to bring out these qualities in their children, while in others gentleness and compassion are given higher priority for both sexes. (Of course, some parents may rebel against their culture, and impart different values to their children.) Teaching the child the difference between good and bad behaviour is only part of the process: myths and stories illustrating moral issues can play an important part. In addition, children learn from example, and from the atmosphere in which they are living.

While the foundations of morality are laid in childhood, parents are not the only influence on moral development. Siblings, other relatives, peers, and many others all contribute. The values of the peer group may differ from those of the family, and some institutions, such as the military, may try deliberately to change the values of their members. Gangs, cliques, secret societies each develop their own variants of morality, with infringements often calling for severe penalties.

It may be helpful here to emphasize again that it is not profitable, and is indeed erroneous, to draw a line between behaviour that is 'innate', 'instinctive', or 'genetically programmed', and that which is due to experience or learning. Most of the behaviour with which we are concerned is based on both—that is, on a propensity to learn some things rather than others. It is the propensities to learn in certain general areas that are properly regarded as part of human nature (in the strict sense, see pp. 10–11), precisely what is learned being honed by individual experience. Humans probably have a propensity to learn preferentially from their parents, and so moral precepts are passed down the generations. Later, learning from peers or authority figures takes precedence.[24,25]

Thus in everyday life individuals are exposed to diverse influences. It has been suggested that we select those that will affect us as the result of three biases. We may simply copy what seems to be common in the group: 'crazes' spread largely because individuals conform to the examples set by others (frequency-dependent bias). This makes sense in that what most people do or think is likely to be the most beneficial thing to do or think in the group's circumstances: it also makes us feel at one with the group (see p. 47). Or we may copy the behaviour, attitudes, or beliefs that we see as most in our own interests (direct bias). This may involve imitating those whom we perceive to resemble us in certain ways, perhaps because what is suitable for them is likely to be suitable also for us. Or we may emulate those whom we perceive to be of high status for one reason or another (indirect bias), for what they do has clearly been good for them.[26] Clearly, more than one of these may affect which moral norms one incorporates into one's self-system (see below). But the extent to which any one of these biases is allowed to operate is affected by the consequences for one's own perceived well-being. Thus we may have an inclination to 'follow the crowd', but that is not always the best thing to do: an individual who aspires to saintliness may perceive it to be in her own interests to follow the example only of those whom she perceives to act prosocially. But if a new norm is in everybody's interests, in the long run natural selection may operate to ensure that it is progressively easier to assimilate so that it becomes part of human nature.

In summary, the acquisition of intuitions about 'correct' behaviour[27] is an inevitable part of development in a social environment. It is due in part to early predispositions to satisfy our own needs, and to act prosocially, shaped in development by parents and others and by the prevailing culture. Parents are in turn influenced in the values they pass on by their circumstances and by the culture. The influences that they and others exert may enhance either prosociality or selfish assertiveness and do so differently in boys and girls.

However, there is no implication that children are consistent in their moral behaviour. For instance, when interacting with their parents they tend to live by norms and values different from those that guide their behaviour with peers. We shall find examples of similar divergences in adult life in later chapters.

Finally, because our moral precepts and values are absorbed in the course of growing and living rather than being taught, we may share them yet differ in how we explain or understand them. Indeed we may not be able to explain why we hold this value rather than that.

THE SELF-SYSTEM AND THE CONSCIENCE

Why do we choose, or refuse, to do the right thing? How is it that individuals choose to align their behaviour with the precepts they have incorporated? This requires a brief digression on the nature of the 'self-system'.

Every individual has an image of the sort of person that she is. Thus one may see oneself as black, well educated, married with two children, a member of the local church, and so on. One has views not only of what one is but also of what values one has. Thus if we are successfully encouraged to be honest in childhood, we incorporate honesty into the ways in which we see ourselves (our 'self-concepts' or 'self-systems').[28] If we have incorporated the relevant norms and values, we behave prosocially automatically in simple situations. In colloquial speech, we act according to our 'moral intuitions' and may not even be able to explain why we behave as we do. Often, however, the situation is not simple, and it is necessary to weigh up the pros and cons by rationally comparing possibilities, though even then we may not act rationally: we shall see examples in later chapters. Thus acting morally always depends on incorporated values, but may or may not depend on conscious decision making.

How one sees oneself is influenced by the context and by whom one is with,[29] and one perceives others through different spectacles from those with which one perceives oneself. For instance, one tends to ascribe one's own actions to the situation, those of others to their personal characteristics. Thus we may say we tripped on a stone, but someone else tripped because they were not looking where they were going. This can have important implications for the way in which we judge other peoples' actions.[30] How often have you said 'I had to have another drink: my host was insistent and it would have been rude to refuse' or 'George drank too much because he is an alcoholic'?

An issue of central importance for the rest of this book is that we like to see ourselves as being consistent, and construct accounts of our lives in which we do not change. We try to maintain *congruency* between the sort of person we see ourselves as being (our self-concept or self-system), how we see ourselves to be behaving, and how we perceive others to perceive us. A person who sees himself as honest, sees himself as behaving honestly, and perceives that others perceive him as honest, would experience congruency. But if such a person is accused of dishonesty, he will try to restore congruency in a variety of ways. He may alter the way in which he sees himself, or denigrate the reliability of his accuser, or behave in a way calculated to change the accuser's opinion. If a person who sees herself as honest realizes that she has behaved dishonestly, she will feel guilt: feelings resulting from such a discrepancy are what we call a bad conscience. To restore congruency she may change her behaviour. Alternatively, she may deceive herself, believing that she was behaving honestly *really*. Or she may find a scapegoat to explain her behaviour.[31] Thus what we call the 'conscience' involves both cognition and emotion. One difference between moral precepts and social conventions is that emotion is less important in infringements of the latter. Of course, what I am describing here is only a simple model of the conscience, but it gives us valuable insight into how we control our behaviour.

It is important to remember that what matters here is one's *perceptions* of oneself, one's actions, and the perceptions of others. What one perceives to be the right course of action may not be right from another's perspective. If one has incorporated a norm of honesty, one may feel that it is justifiable to say hurtful things to someone else in order to live up to one's principles. One may even feel impelled to act against one's own interests by confessing a misdemeanour that would otherwise be of no significance. In general, it is *feeling* that this is good and that

25

bad which matters, though these feelings may be engendered by what we have previously learnt about the consequences of actions. We do the right thing, often without conscious thought, to maintain congruency in our self-systems. If we do the wrong thing, and lack congruency, we feel guilt. Although we may act without conscious thought, the outcome may be influenced by unconscious evaluations of the context and our emotional state.

When one acts knowingly in a way that is contrary to one's moral intuitions, deceiving oneself can take the form of justifying one's action with another moral precept—for instance during World War II a friend smuggled watches through Customs when returning from South Africa, where watches were cheap, to the UK, where they were almost unobtainable. He justified lying to Customs by telling himself that lying was alright in this case because it was his duty to use this opportunity to relieve the shortage. It is noteworthy that in this case the excuse involves the use of a moral rule valid in other contexts. Presumably, the more readily such alternatives are available, the more likely one is to behave contrary to the generally accepted standards.

One may ask whether a perfect society in which nobody broke the rules would be possible. There are reasons to think otherwise. We shall see later (pp. 36–9) that natural and cultural evolution have ensured that individuals have propensities both to cooperate and to compete with other members of their group. Individuals who were consistently prosocial cooperators might lose out in competition with others who were better at competing. The circumstances of our lives force us to be both cooperators and competitors, prosocial as well as antisocial. Our consciences keep us on the right track, but we have ways of circumventing our consciences when we stray.

A moment's introspection forces one to admit that one does use different standards in different contexts. For instance, friendship relationships involve a norm of caring and responsibility that is at least less conspicuous in interactions with strangers.[32] In general, the differences in the norms we live by in our several relationships are dictated by the culture, but are usually also such as to serve our own interests. Usually, but not always. The norms and values that we acquire may sometimes be to our own detriment—sacrificing one's life to save the life of a stranger would be an extreme example, cigarette smoking because it is fashionable would be another.[33]

The norms incorporated in our self-systems affect not only our own behaviour, but also the way we respond to that of others. If we see someone breaking the rules by which we try to live, we feel righteous indignation and may seek to intervene. The importance of this will become apparent in the next chapter.

CONCLUSION

To summarize, we have propensities to behave helpfully and cooperatively with others (Prosociality) as well as to do the best for ourselves (Selfish assertiveness). Moral codes maintain an appropriate balance between these two propensities. They involve *principles*, which are pan-cultural, and *precepts*, specific to one or a number of cultures. The principles are basically a product of our evolutionary past, and can reasonably be presumed to result from natural selection. The precepts have been affected over time by two-way relations between what individuals do and what they are supposed to do in the culture and environment in which they live: how people behave and what people value affects the culture and its moral precepts, and the moral precepts affect people's behaviour and values. In general, precepts are compatible with the principles. The development of morality within the individual involves both basic propensities and the acquisition by individuals of values and precepts from experience with parents and other members of the community. Moral behaviour depends on perceived congruency between what one sees as the right thing to do and what one sees oneself as doing. We strive to maintain congruency between what we perceive ourselves to be doing, what we perceive to be right, and how we perceive others to perceive us. People can perceive themselves to be acting correctly when others think differently.

2

The Evolution of Morality

The last chapter contained a brief summary of how we gradually assimilate the rules by which we live as we grow up. The moral choices we make, the precepts that govern our behaviour, and the virtues we admire are influenced by the rules incorporated in our personalities as the result of earlier experience. While we feel that we 'should' act according to the moral principles and precepts that we have acquired,[1] this begs the question how it has come about that this behaviour and this virtue are to be admired and emulated, and not those. There are really two questions here, one biological/cultural and the other sociological/psychological. The first, the subject of this chapter, concerns how natural selection gave rise to the moral principles that are common to all cultures (human nature in the strict sense). The second, asking about how moral rules change and differ between societies, is related to the evolution of legal systems, and is discussed in that context (Chapter 3).

THE EARLY EVOLUTION OF MORALITY

Bernard Williams, one of the most respected of modern moral philosophers, has written that 'the world to which ethical thought now applies is irreversibly different, not only from the ancient world but from any world in which human beings have tried to live and have used ethical concepts'.[2] Whilst salutary in some contexts, Bernard Williams's view must not be taken to imply that an evolutionary approach can tell us nothing about ethics today. Biological evolution moves slowly, and much of our behaviour is still guided by principles that served us well earlier in our evolutionary history. This includes prosocial as well as antisocial behaviour. But seeking for the evolutionary sources of morality is not the same as saying that what is natural is good: propensities that promoted survival and reproduction may have been seen as acceptable earlier in human history but may not be seen as good now.

We have seen that potentials for prosociality and selfish assertiveness are part of human nature, but how has that come about? Evolution involves selection for characteristics conducive to the survival and reproduction of individuals. If sharp eyes help you to find food, the individuals who have the sharpest eyes are most likely to survive and breed, and their offspring are likely to have sharp eyes too. In the same way, it is easy enough to see how a propensity to look after one's own interests evolved: selfishly assertive individuals are likely to win out in competition with their peers and leave more offspring with similar genetic constitutions. But how could selection favour behaviour conducive to the well-being of others? Surely individuals who behaved selfishly to others would have better access to resources and prosocial individuals would be eliminated? There are two cases to be considered here: prosociality to genetically related individuals, and prosociality to individuals who are not closely genetically related.[3] I shall also consider the origins of other aspects of morality: behaviour to social superiors, relations between the sexes, and behaviour to other group members.

RELATIONSHIPS BETWEEN KIN

We see it as natural that parents should make sacrifices to look after their children, and that children should respect their parents. This is not surprising, for such behaviour is in keeping with the mode of action of natural selection. Natural selection operates to ensure that individuals behave in a way that maximizes their lifetime reproductive success and also that of their descendants and close relatives. The genetic constitution of those who leave more healthy offspring than others will be better represented in succeeding generations. In looking after their children, parents are ensuring the survival of individuals who behave in much the same way as they do and are likely to pass on a similar complement of genes. Thus looking after one's children is in keeping with the dictates of natural selection.

Clearly children share genes with their parents to the same extent as parents share genes with them. Children who look after their parents are ensuring that individuals with a similar genetic constitution to themselves are better able to have further children and, pragmatically, such children are also increasing the disposition of their parents to look after them. By the same token, children in a family are closely related genetically to each other, and are predisposed to behave prosocially to their siblings more than they are to strangers, and to a lesser degree to other relatives, aunts, cousins, and so on, according to their degree of relatedness. This principle of 'kin selection' applies throughout the whole of the animal kingdom.[4]

Thus our moral indignation when a parent neglects a child, or when children do not respect their parents or cherish their siblings or behave with proper respect to their aunts, is in keeping with the forces of natural selection—aided no doubt by culturally selected precepts such as the Old Testament 'Honour thy father and mother that thy days may be long in the land that the Lord thy God giveth thee'. It is incidentally worth noting that this precept, like many others, is backed up by an appeal to individual selfish assertiveness.

But relationships between parent and child are not all sweetness and light. Some degree of conflict is nearly always present.[5] Babies want to be fed when parents want to sleep. The growing desire for autonomy may lead the child to displease the parent. In addition we feel that there are limits to the extent to which parents should look after their children: thus parents may be considered as 'over-protective', children as 'over-mothered'. In other words, our views about 'correct' behaviour change with the age of the child. As children grow older, most parents feel that their responsibilities towards them diminish. This is in part a consequence of the child's assertiveness leading to increasing demands for autonomy: 'I should be allowed to stay out as late as I like', or 'to wear whatever takes my fancy'. How is this to be explained?

Much such conflict has its roots in the forces of selection, and involves not only a conflict of interests in the here and now, but also conflict in an evolutionary sense. The biological issues involved are by no means peculiar to human beings, and once again such conflicts are in keeping with the theory of natural selection. This is because it is in the (biological) interests of parents to look after their children only so long as doing so does not overly impair their own prospects for future reproduction. But growing children continue to make demands on their parents until they demand more than their parents are prepared to give. Conflict is inevitable. Indeed, conflict starts before birth, the foetus demanding more than the mother can give, and this may lead to a variety of complications such as gestational diabetes. This biological approach, supported by much data from other species,[6] also provides a perspective on conflict between siblings, who are likely to compete for parental attention, but support each other in disputes with outsiders. In addition, a later-born must have priority over an older sibling in many aspects of parental care, with a resultant decrease in parental attention to the older child: this is a potent cause of 'sibling rivalry'. A last-born is especially likely to get good parental care, as there are no other children for whom the parents should conserve resources.[7]

Although I have placed emphasis on long-term biological or cultural factors in conflicts between relatives, this should not be taken as indicating that proximate factors are unimportant. Of course, every disagreement has proximate causes. And the fact that they have evolutionary roots does not mean that such conflicts are insoluble. Awareness of our biological Achilles heels may help to minimize problems.

Some other cases that seem to be exceptions to the general rule of prosociality between parents, children, and other family members, on closer inspection appear to be exceptions that go some way towards proving the rule.

First, the practice of contraception reduces the number of children born and seems contrary to the principle of natural selection. However, from a biological point of view, it is always necessary for parents to balance the needs of children already born against the demands that would be made on their resources by further conceptions. It may be more biologically advantageous to leave a few healthy children who are likely to reproduce themselves, than a larger number of less healthy children whose reproductive potential is more limited. In many societies lactational amenorrhoea and cultural devices such as a post-partum sex taboo facilitate this. In modern societies the issues may be different, with women wanting greater autonomy and parents desiring a lifestyle different from that which would be possible with children. But there is another issue that is becoming increasingly important. In an overpopulated world, many are beginning to feel that it is morally right to limit one's reproduction. What is natural, in this case to have many children, is not necessarily right.

A second exception concerns the incidence of infanticide and child abuse. This seems contrary to the evolutionary principle that parents should look after their children. However, the evidence shows that most cases of infanticide or child abuse involve either (a) Step-parents, who have no close genetic relationship to the child, or (b) Mothers who are young, undernourished, or ill, and thus unlikely to raise the child successfully and would do better to husband their resources, or (c) Infants who are ill or handicapped in ways that make their eventual reproduction improbable.[8]

A third exception concerns the incidence of adoption. In nonindustrial societies this usually involves adoption by relatives who have some 'genetic interest' in the child.[9] In other cases the adoption may be perceived by the parents to be in the child's interests as a sort of apprenticeship. In industrial societies the adopting parents are often people who cannot otherwise satisfy their desire for parenthood, itself partially biologically determined.

Thus many of the properties of relationships between genetically related individuals can be understood in terms of natural selection, though their dynamics are affected also by the demands of the current situation. It is *perceived* relatedness that is important, not always absolute genetic relatedness. This conclusion is supported by experimental work showing that when confronted with potential life or death situations, people's decisions as to whom should be saved are usually those that would be predicted by the theory of natural selection. Specifically, humans are favoured over other species, kin over other humans, close kin over distant kin, and friends over strangers. In the last case, familiarity is presumably perceived unconsciously as indicating likely relatedness (see also p. 35). In less life-threatening situations there is a tendency to favour the young and old over those of intermediate age, the sick over the healthy, the poor over the rich, and women over men.[10] In these latter cases precepts of charity and politeness are presumably operating.

RELATIONSHIPS WITH NON-KIN

Prosociality to more distantly related individuals can not easily be accounted for in terms of natural selection, and there has been much debate over how it could have arisen: it might seem that those who look after their own interests are bound to do better in the long run and natural selection would act against helping non-relatives.

One answer is that it may pay to help or cooperate with another who is likely to reciprocate at a later date. This occurs rarely in animals,[11] and its generality is controversial. Reciprocity is certainly an important issue in human behaviour, as discussed in the next chapter. But simple reciprocity is unlikely to have been important in promoting the biological evolution or maintenance of prosocial behaviour in large human groups, where the chances of meeting the same individual again are relatively low.

A more important issue probably lies in the importance of living in a group that is in competition with other groups. Early in human history, individual survival depended on membership of a cohesive group, and even today we function better if living in a harmonious society, where individuals cooperate with each other, than in one composed of selfish individuals. It is reasonable to suppose that early human groups competed with each other (either directly, or by being better at acquiring limited resources), and that groups with a higher proportion of individuals who behaved prosocially to, and cooperatively with, other in-group members tended to do better than those containing many self-ishly assertive individuals. It is therefore necessary to consider competition and cooperation between individuals within groups and also competition between groups.

Natural selection acts on genetic differences between individuals. Within each group, those individuals who looked after their own interests would be likely to be the more successful and leave more offspring, and tendencies for selfish assertiveness would be selected for. But competition between groups is likely to depend on prosociality, including cooperation and reciprocity, between in-group members. Could prosociality have been genetically selected for through its value in competition between groups? Evidence from animals makes this seem unlikely, for genetic evolution by selection between groups has been at best rare in animals.

But humans have cognitive abilities not available to other species. We must therefore ask, in human groups could *culturally* maintained similarities between individuals within groups and *cultural* differences between groups provide a basis for *group* selection?[12] Could, for instance, a group consisting mainly of prosocial individuals not only be maintained through successive generations but also succeed in competition with groups consisting mainly of selfish individuals? Could a high proportion of prosocial individuals enable the group to perform better in cooperative enterprises, to share knowledge, or to be more effective in combat? Computer modelling shows that prosociality to other group members could bring a group advantage if certain conditions were fulfilled—namely if naïve individuals tended to conform to the behaviour that was most common in the group, if the environments in which the groups lived were heterogeneous, and if the groups moved from time to time.[13] All these conditions are likely to have been fulfilled early in human evolution. Thus uniformity within groups, and diversity between groups, provides conditions that could favour prosocial behaviour to fellow group members. Although the precise selective processes that have operated are not yet universally agreed, both selfishly assertive and prosocial behaviour could have arisen through selection, the main emphasis being on genetic selection in the first case and cultural selection in the second.

But any selfish individuals in the group would still do better in competition with other group members, especially if they interacted mostly with prosocial individuals. Only if prosocial individuals interacted preferentially with other prosocial individuals could they increase in frequency in the group. This could happen if prosocial individuals could distinguish between prosocial and selfishly assertive individuals, and this would be the more important, the higher the proportion of selfish individuals. One obvious clue to whether an interaction with another individual will lead to one's own benefit is the way the other initiates it: only if the other seems prosocial or cooperative or honest is it worthwhile proceeding with the interaction. Thus the persistence of *prosocial reciprocity*, as implied in the Golden Rule, requires a prosocial initiation.[14]

The propensity to act prosocially is supported by another mechanism, 'the moralistic enforcement of norms'. In the first place, experimental evidence indicates that people are especially adept at detecting the breaking of a rule if the rule involves a social contract. The classic experiment involves the contrast between an abstract problem and a social problem. In the former, the experimenter uses a pack of cards with a letter on one side and a figure on the other. He lays out four cards, thus:

D   F   3   7

He also tells the subjects that there is a rule that a card marked D on one side must have a 3 on the other, thus: the students must say which cards must be turned over to decide if the rule is false.

The social rule problem is similar, the rule being 'You must be over 21 years old if you are drinking beer' and the cards showing an individual's age on one side and what he or she is drinking on the other.

20 years   Beer   24 years   Coca-Cola

Most people find the first problem much more difficult than the second (the answers are given in the notes).[15] Although the results of such experiments depend on the conditions and wording, they suggest that humans are adapted to understand social contracts and to detect violations to them.[16]

In addition, experimental evidence indicates that we have (or develop) a special facility for recognizing those who have previously been labelled as cheats.[17] Individuals who cheat on their social obligations are punished by others, while, perhaps even more importantly, individuals who behave morally receive approval from others.[18] Presumably individuals benefit by monitoring the behaviour of others and assessing them as trustworthy or not on the basis of their observations, because they can then choose with whom to interact. Gossip plays a useful part in disseminating their conclusions. Receiving moral disapprobation is likely to decrease an individual's access to resources or chances of breeding successfully, moral approbation to increase them. Punishment may involve expulsion from the group, public humiliation, threat of divine punishment, or physical violence. Of course the act of punishing imposes some costs also on the punisher, but data from experimental games indicate that individuals are in fact willing to incur costs in order to punish a wrongdoer. In the long term, these costs are presumably more than compensated for by the beneficial effects on the group and thereby on the individual.[19]

How far the tendency to punish immoral behaviour depends on experience is unknown: it has been suggested that the righteous indignation that leads to punishing non-cooperaters even when we are not personally involved could have arisen from the doubt that observing such an act throws on our own moral code, and the consequent lack of congruence[20] (see pp. 25–7).

A number of aspects of human behaviour follow from this. Such moralistic enforcement is more likely to be effective if people tell each other about the pro- and antisocial behaviour of others—that is, if people gossip about how others behave.[21] However, the possibility of being punished for breaking the rules encourages people to appear to be more public-spirited than they are:[22] it is probably as a result of this that humans have become adept both at deception and at detecting the breaking of a social rule.[23] It has been suggested that, in modern societies, the direct effects of the damage to reputation resulting from immoral behaviour may be more significant to the individual than its indirect effects through diminishing the integrity of the group.[24]

Modelling techniques show that moralistic strategies that involve cooperating, punishing those who do not cooperate, and punishing those who do not punish non-cooperaters, could be stable. Provided the cost of punishing is not too great, a few punishers in a group can maintain a norm of cooperation (or any other norm).[25] Thus both competition with in-group members and cooperation with them may lead to greater reproductive success. Part of the thesis presented here is that morality serves to maintain an appropriate balance between them.

As noted above, we know surprisingly little about the early ontogeny of many of the qualities important in social interaction, but it is reasonable to suppose that the precise course of learning, based on pre-existing propensities, is influenced and to a large degree determined by the customs and values current in the culture. This may occur because prosocial behaviour induces parental approval, or because of the tangible rewards that it brings, or the public recognition that results. As an example of the last, it is good to be perceived as trustworthy, because profitable exchanges may follow (though of course this opens the way for pretence).[26] As another example from a quite different context, much controversy has centred around the question of why hunter-gatherers should expend so much energy on hunting large animals. They provide much more meat than can be consumed by the hunter or his family, and better returns can be obtained from going after the more easily acquired small game. The evidence points to the view that great kudos attaches to the killing of a large animal, and this brings later dividends in terms of attractiveness to women and warning off other males.[27]

SELFISH ASSERTIVENESS, STATUS, AND THE 'VIRTUE' OF HUMILITY

We have seen selfish assertiveness as a source of behaviour that may be interpreted as 'bad' by parents in infancy, and as a source of selfish autonomy in later years. And I have written as though the main purpose of morality is to hold it in check. But one must not be too hasty. Some degree of selfish assertiveness is essential for individual survival, and what matters is the balance. Even to enter a relationship often requires at some level a degree of selfish assertiveness, though profitable interactions with others[28] would not be possible without the virtues stemming from prosociality (pp. 37–9, 88).

In probably every society, selfish assertiveness leads to individuals seeking status. Some small-scale human groups appear to be egalitarian, but that is because the assertiveness of individuals is held in check by others and by social norms: the propensity for selfish assertiveness is present but suppressed if it gets beyond interpersonal disputes. Similarly today selfish assertiveness in the form of boasting is looked at askance, at least in the UK. Nevertheless, in most societies status differences exist. They may be based on a variety of attributes, including physical prowess, beauty, wealth, wisdom, or generosity, that benefit (or are envied by) the rest of the community. High status usually brings both responsibilities and rewards: presumably the latter are perceived as more than outweighing the former so that status is sought as a valued resource in its own right.

It may be in the interests of those lower in status not continually to strive for higher status but to accept their lot. Indeed high status is not always desirable as there are bound to be costs in maintaining it. Those lower down may already gain from their dominance to those even lower than themselves, so that an attempt to climb higher would not be worthwhile. Furthermore they may gain from the efforts of high status individuals to maintain their positions by providing benefits to those below them, for instance by protecting them.

In addition, lower ranking individuals may be encouraged to be content with their status by the values of the culture. These values arise because those having high status can formulate rules and moral precepts that are to their own advantage. For instance, 'humility' has been part of Christian morality, but the Church of England's Catechism requiring the confirmand to undertake 'to order myself lowly and reverently to all my betters' helps to maintain not only the hierarchical structure of society, which may or may not be for the good of all concerned, but also the status of the rule-makers. How much the rule-makers were selfishly motivated and how much they had the good of society at heart are open issues. The result has been to make many feel that it is right to be humble, that it is wrong to question authority. In this context, it is also of interest to note that the Hebraic Commandment concerned with the relations between parents and children

(p. 32) insists in children honouring parents and not vice versa: presumably it was written by parents.

It is also in the interests of influential individuals to ensure that precepts and values conducive to the maintenance of the status quo are seen as absolute. The purveying of religious systems in most societies has been the responsibility of the religious leaders, who make moral pronouncements for which they claim divine authority. Secular leaders also often claim divine authority to lend force to their decisions. In insisting on the absolute nature of the precepts that they purvey, they can threaten the transgressor with dire punishment in this life or the next.

Once status relationships have been established, the force of authority can be so powerful that it can be used to cause individuals to act in ways contrary to their moral beliefs. An experimenter, perceived by the subjects as an authority figure, could cause experimental subjects to administer what they believed to be strong (but were actually bogus) electric shocks to a confederate of the experimenter. The subjects continued to obey the experimenter's orders to increase the strength of the shocks even though they believed the confederate to be in great pain.[29]

The influence of powerful individuals may be more subtle than that. Take the conventional harming versus not helping problem. Suppose a doctor could save five ill patients if he killed a sixth healthy patient and distributed his healthy organs amongst the sick ones. Should he do so? One person would die but five would live. Most people opt for doing nothing, apparently because they see killing one individual to be more heinous than failing to save five.[30] Harman[31] has suggested an explanation for this. All individuals in a society would be better off if everybody desisted from harming others. But if there were a suggestion that everyone should help others in need, the powerful would be called on to help more than those less well provided for. Perhaps they have influenced the two-way relations between what people do and what they are supposed to do so as to empower 'not harming'. (The related question of human rights is discussed in the next chapter.)

SEX AND GENDER-RELATED ISSUES

Every culture has norms, usually seen as moral precepts or laws, regulating relations between the sexes. In none is total promiscuity allowed, but some cultures are monogamous, some polygamous, and a few polyandrous. However, in at least the majority of societies the rules governing sexual behaviour differ between men and women. There are parallels between these differences and those between the reproductive requirements of men and women and between differences in their anatomy and physiology. These differences can be seen as direct consequences of the action of natural selection acting through reproductive success. The biological arguments here apply to virtually all mammals as well as humans. A few examples follow.

While the lifetime reproductive success of a male is potentially almost unlimited, that of females is limited by the time necessary for pregnancy and lactation. Competition between males for females is therefore stronger than that between females for males, and males tend to be larger and stronger. Male aggressiveness and assertiveness have therefore been selected for. Machismo traditions and protectiveness towards women are in harmony with this, for men not only compete for mates, but also must protect their mates from the attentions of other men.

A female knows that the child in her womb is her own, but a male can be cuckolded. Thus extra-pair mating by a female in a monogamous or polygynous society may result in the male expending parental care on infants that he has not fathered. By contrast, extra-pair mating by a man involves little material disadvantage to his wife, providing he does not expend their resources on other women. In harmony with this, men are allowed more sexual licence than women in virtually all, and perhaps all, societies. Women are expected to be chaste and faithful. This has been institutionalized in many societies, and lies at the base of the Muslim harem and the linkage between male machismo and female virtue in many Catholic cultures. This does not mean that male promiscuity is necessarily seen as right. In a culture that values monogamy and fidelity, male infidelity may have a variety of undesirable consequences, including psychological damage to the wife and children. As has been emphasized already, what is natural is not necessarily right.

Third, not only may a male have been cuckolded, but also he needs to expend fewer resources in creating another infant than must a woman. Thus it makes biological sense that offspring tend to be valued more by their mothers than by their fathers.

Fourth, it is likely that, earlier in human history, a mother was more likely to rear her offspring successfully if she had support from a partner. It is thus to be expected that a secure relationship with a partner is more important to women of childbearing age than to men. If the partner were male he would be likely to give more effective support than a female as he would not be concerned with a child of his own; and the male who had fathered the infant would have more biological incentive for protecting its mother than would another male.

These and other parallels between how people actually behave and biological predictions are in keeping with the view that many cultural stereotypes and expectations have arisen as a result of biological predispositions, and have been honed by the reciprocal influences over time between what people do and what they are supposed to do in the culture and environment in which they are living. Although the basic differences in behavioural predispositions between men and women may be small, members of both sexes seek to behave in ways that have brought reproductive success to their predecessors and are now seen as appropriate for their gender. They will also support the norms for the other sex if they are in keeping with their own interests. It is in a man's interests that his partner should be nurturing and caring, and in a woman's interests that her partner should be a good provider and successful in competition with other men.

The origins of the cultural differences in the relations between men and women have been little studied, principally because of the paucity of the evidence. But one matter has received a good deal of attention, namely the ubiquity and diversity of incest taboos. To the biologist, incest means mating between biologically closely related individuals. Considerable evidence shows that close inbreeding can be deleterious to reproductive success in both animals and humans.[32] This is partly because many infectious agents can mutate rapidly in a way that makes it possible for them to remain undetected by the immune system of their host. Out-breeding may provide sufficient genetic distinctiveness for the immune system to be able to recognize, and deal with, a broader spectrum of infectious agents.[33] Thus the existence of incest taboos has an evolutionary explanation. There is indeed some evidence for a mechanism that would help to reduce inbreeding in humans, namely a biological inhibition against mating with someone with whom one has been familiar since early childhood: familiarity serves as an indicator of genetic relationship.

The diversity of the incest rules amongst human groups, however, requires further explanation. Anthropologists and other social scientists, impressed by the diversity of incest taboos and by the fact that the prohibitions often do not concern close relatives, define incest as sexual relations between those whose cultural relationship debars them from a sexual relationship. Although close kin are nearly always included in the taboo, there are a few societies, probably because of local circumstances, that permit and even encourage it. But in many pre-industrial societies, marriage to a woman from another group is obligatory, and forms the basis for aggressive raiding. In other cases the prohibitions may be in the interests of powerful individuals or groups. In Europe the Christian Church discouraged close marriages, thereby weakening consanguineous ties and increasing the Church's power to obtain bequests.[34]

GROUP LOYALTY

A final issue here, of central importance to most of the later chapters, concerns the felt importance of belonging to one or more groups. Humans have always lived in groups—originally probably mainly limited to close relatives, but subsequently elaborated into enormous, complexly organized societies. As noted earlier, living in a well-integrated group was advantageous to group members in furthering group endeavours, especially those in which numbers counted (at least up to a point).

The evolution of the many traits that enable us to live in groups is not an issue that can be pursued in any detail here,[35] but it can be seen as an extension of prosociality to non-kin (pp. 35–40). We have already seen that individual success has depended both on a degree of self-assertiveness in competition with other in-group members, and on prosociality to them, the latter leading to group cohesion and to the success of the group in competition with other groups. Group loyalty has become objectified as a moral right as a consequence of this conflict. Being loyal has become part of the way a good group member sees himself, that is, part of his self-concept or self-system (pp. 24–7).

There is an important issue here. The way that individuals see themselves includes not only their personal characteristics, but also their group membership. Their self-systems include both an individual identity and a social identity: a woman may see herself not only as a lawyer, a mother, virtuous, and so on, but also as a member of a particular church or sports club. Members of a group see themselves as a group, and as more similar to each other than to outsiders on characteristics significant for the way in which they define the group. They are likely to see themselves as interdependent: the more they see themselves as interdependent, the more cohesive the group; and the more cohesive the group, the more they see themselves as interdependent.[36] Attraction to others perceived as similar to oneself is an important contributor to group cohesiveness: this is especially potent if the perceived similarity involves unverifiable attitudes and beliefs.[37] Religious systems are an obvious example here, and can play a central role in maintaining group loyalty: religious differences from a rival group have often been exploited by leaders to ensure both loyalty amongst their own followers and antagonism to members of an out-group. There is no need to point out that this has been a major cause of human misery.

It has been suggested that the degree of cooperation with in-group members depends on the perceived stability of the group. If group stability is threatened, it will pay individuals to serve the group and maintain its stability. However, if the group is perceived to be secure, individuals can look after their own interests. There is some evidence for this view in extreme situations: prosociality to in-group members in the form of patriotism comes to the fore in times of war.[38]

Members of a group tend to elaborate group values, behavioural norms, and explanations of events. The values and norms that emerge tend to be such as justify the individual's and group's goals and facilitate achieving them. Furthermore, individuals elaborate markers that distinguish their own group from neighbouring ones. It is probable that many cultural characteristics

47

arose in this way—the Jewish and Islamic dietary restrictions being a case in point. Dialect and language differences also act in this way. Such markers are often seen as basic and unchangeable.

That competition between groups played an important part in human history means that prosociality has been selected for or encouraged only towards in-group members. Success in competition between groups depends on loyalty and duty to the group to which one belongs, but not to outsiders. Because it is coupled with opposition or antagonism to out-groups, it provides a basis for xenophobia. We shall see in later chapters that flexibility in the perceived boundaries of the in-group is an important way of bending the rules.

EXTENSION OF THE IN-GROUP

While for a more harmonious group, prosociality to members of one's own group clearly needs to be emphasized, for a more harmonious world, prosociality needs to spread beyond any limited in-group. In practice this has been happening for some time, though too slowly. Early in the nineteenth century in the UK a pauper could obtain relief only from his own parish, though communal disasters, such as the famines associated with the Highland clearances, sometimes received relief from outside. Towards the end of that century the relief of poverty in fellow citizens had become a national concern. Now we send food, medical and educational aid to distant parts of the world. Globalization, although a two-edged sword, could further this process. But aid to other countries or distant individuals generally occurs only if it is seen to be advantageous to, or at least not to the detriment of, the donor country or individual. Pragmatic reasons may be necessary, but hopefully ethical considerations may come to suffice—the more so if they bring a feeling of righteous satisfaction to the donor.

Once one sees oneself as a member of a group, prosociality to in-group members enhances the congruency of one's self-system (pp. 25–7). The feeling that one should act prosocially is stronger if the behaviour is to be directed to a member of one's own group rather than an outsider. The criteria by which one recognizes fellow group members is thus crucial. If I feel that it is important to be kind to friends, I may also feel that I should be kind to strangers; and if I should be kind to other members of my group, I should also (though perhaps less strongly) be kind to foreigners because they share my humanity. And (though perhaps even less strongly) be kind to animals because they also are sentient beings.

CULTURAL DIFFERENCES IN MORALITY

Thus there are basic commonalities in moral codes that are derived from commonalities in human nature and are essential for successful group-living in any environment. A principle of prosociality to group members is ubiquitous in all societies, but the way in which it has been elaborated into precepts has come to differ between cultures. How does this affect our view of the behaviour of members of other cultures?

We regard the moral code of another culture as wrong if it contravenes the Golden Rule. Rarely the precepts of a culture contravene the principle of prosocial reciprocity because other precepts or rules of conduct have been imposed by a leader: we must then judge *the moral code of that society* adversely. This can happen because the leader sees (or claims to see) the persecution of one group within the society to be to the advantage of his own position or the society as a whole: the persecution of Jews and Gypsies in Nazi Germany is a recent example.

The *precepts* that have been derived from the basic principles in the course of reciprocal influences between culture, behaviour, and environment within one culture may differ from those in another. Unless we have full understanding of the history and circumstances of another culture, we have no certain basis for regarding the behaviour of an individual from another culture who conforms with its precepts as wrong just because his behaviour is not in keeping with the precepts of our own culture. We may still regard an individual who contravenes the principle of prosociality to be culpable if we judge that the principle has not been drummed out of him.

SUMMARY AND CONCLUSION

The general structure of moral codes can be seen as the result of natural and cultural selection. It serves to maintain an appropriate balance between prosociality and selfish assertiveness in relations between individuals in small groups.

Selfish assertiveness can be accounted for as the direct result of natural selection, those who look after themselves tending to leave more descendants.

Prosociality directed preferentially to kin is largely a result of natural selection, as it benefits individuals likely to acquire and pass on some of the actor's genes.

Prosociality to non-kin is less easy to explain, as the actor is behaving prosocially to potential competitors. The evidence indicates that it is largely the result of cultural selection between competing groups, those groups that contained the greater proportion of prosocial cooperating individuals being more likely to succeed in competition with other groups for limited resources than those groups containing fewer.

Some virtues, like humility, may result partly from leaders purveying precepts in their own interests.

All cultures have precepts governing sex and gender-related issues, including restrictions on those with whom sexual relations are permitted, but the rules differ between groups.

Cultural group selection can promote prosociality only towards in-group members. Hence the universal tendency to categorize others into members of the in-group or out-group and to regard loyalty to the in-group as a moral issue. We shall see later that this in-group loyalty often contributes to actions that would not otherwise be accepted in the society in question.

Cultural selection has honed the moral precepts so that they differ somewhat between groups.

3

Ethics and Law

We have seen that the evidence indicates that morality evolved within the context of interpersonal relationships in small, relatively simple groups. One source of evidence on the early elaboration of moral precepts comes from the early history of legal systems, for laws were originally elaborated and formalized from moral intuitions or principles as an additional way to control the behaviour of individuals. This chapter, therefore, contains first some speculations on the early evolution of moral systems, and then specifies some of the similarities and differences between ethics and laws.

THE EMERGENCE OF MORAL AND LEGAL SYSTEMS

We can know little about the early emergence of moral and legal systems because the early stages did not produce written records. However, such evidence as there is suggests a close relationship between law and moral precepts.[1] Because this evidence is mostly indirect and may involve special cases, much of the brief summary that follows involves diverse and unrelated sources of evidence brought together into a possible scenario, and must be regarded as highly speculative.

Modern hunter-gatherers tend to live in marginal situations in environments ranging from tropical forests to Polar deserts, and there can be no guarantee that their social arrangements represent an early stage in the development of modern societies. But such evidence is nearly all we have, and suggests that early hunter-gatherers probably lived in small family groups, though inter-group relations probably led to individuals having 'social knowledge' of more than a hundred others. Groups consisted of individuals most of whom were related to each other, and group-living presumably provided protection and was valuable in cooperative hunting.[2]

In most modern hunter-gatherer groups, help is given to band members who are ill, incapacitated, or ageing, provided that such help does not threaten the overall welfare of group members. Of course, within the group individuals seek to satisfy their own needs: males compete for women and both sexes compete for resources, but the groups tend to be egalitarian. Too much selfish assertiveness is seen by others as threatening their own interests, so that individuals gang up to suppress any tendency to self-aggrandizement. A tendency to suppress individuals who attempt to dominate has been described even in groups who have abandoned the hunter-gatherer lifestyle; and in some quite complex societies, such as the North American Blackfeet, leadership was muted, with decisions taken collectively.

Some conflicts are settled simply by the parties separating or by the offending individual being ejected from the group. However, peaceful relations within each group of hunter-gatherers is ensured partly by gossip about those who infringe group norms, but ultimately by the readiness of individuals or their kin for revenge: this provides individuals with a defence against attack or exploitation. Revenge has a long history: stories about revenge are to be found in the earliest known literature, such as the Legend of Gilgamesh. It has also been described in the twentieth century in some pre-industrial societies. For instance, after a homicide amongst the Nuer vengeance was the most binding obligation of paternal kinship: homicide was likely to lead to a blood feud, which might then influence relations between larger groups which participated indirectly in the conflict.[3] Blood feuds carrying over successive generations persisted until recently even in parts of the Western world, such as Albania.

But revenge can lead to escalation for four reasons. First, it may not be in the same currency as the original offence, with what is seen as appropriate often depending on the relative status of the participants: it is then difficult to judge what is fair. Second, in any case, it is what is *perceived* to be fair that matters, and injury committed against oneself or one's kin is likely to be perceived as greater than the injury one has inflicted on others. Third, in a prolonged exchange, the causes of one's own actions and the consequences of the opponent's actions are more salient than the causes of the other's actions and the consequences of one's own. And fourth, an element of righteous anger may add to the retribution inflicted.[4] Escalation is especially likely if rival kin groups are involved.

Revenge can be seen as a form of reciprocity (pp. 84–5), and must have been based on shared understandings about what behaviour was and was not acceptable. Presumably these shared understandings were the results of the need for balance between selfish assertiveness and prosociality. Where property was involved, a concept of possession or ownership is implied. In due course these understandings became formalized. This could have been facilitated by the tendency to conformism, which we have already seen as likely to have been characteristic of early human groups (see pp. 36–7). If successful groups were those with high levels of prosociality and cooperation, conformism could have led to what most people do becoming what they were expected to do.

The formalization of precepts may have been the joint result of the experience of individuals and/or the insistence of charismatic Moses-like figures, motivated either by their own wishes or the good of the group. It is thus likely that such rules involved the expression both of the views of leaders and of views already communally accepted. Once formalized, punishment could have been applied for non-compliance and, by mechanisms similar to those that operate today, those who tried to prevent or punish antisocial behaviour in others could have been rewarded.

As leaders emerged, it would then have been in their interests to maintain their status within the group, and to promulgate rules/laws to preserve harmony and the status quo (see pp. 40–2). A well-known example in the second millenium BCE is the emperor Hammurabi who claimed divine guidance to validate his authority.[5] Leaders would have been likely to promote values that were in their own interests, such as the moral value of obedience to their authority. However, leaders who did not possess absolute power and who went too far in trying to exploit their positions for their own benefit were likely to have their power limited. In this, cooperation between sub-leaders or the group as a whole probably played a role, as in the events that led to Magna Carta. In addition, reciprocity may have operated, leaders retaining respect only if they themselves provided some service, such as protection or access to resources.

The formalization of laws involved primarily restrictions on what people should be allowed to do, and not prescriptions on what they should do. However, the integrity of these groups cannot have depended solely on punishing the perpetrators of disruptive behaviour. Prosociality and some tendency for cooperation must have existed if the groups were viable in the first place: since the groups consisted largely of related individuals, this could have arisen through kin selection (see pp. 31–5). Gossip about free-riders may have been adequate to prevent most antisocial behaviour and to ensure cooperation on group enterprises. In addition, because the well-being of individuals depended on that of the group, and vice versa, all group members had an interest in the prosociality of others. As a result, those who behaved prosocially would have been liked and admired. That would have brought status, leading to greater success in everyday life and in reproduction.

So far I have assumed a simple social structure. However, as successful groups grew larger they must have developed into larger entities with sub-units held together by a common language, religion, or rituals, and coming together from time to time.[6] This was the case, for instance, amongst the Australian Aborigines. Amicable relations between bands or groups may have been maintained by ritual and/or the exchange of gifts. The moral codes of such groups, while constrained by the basic principle of prosocial reciprocity, must have diverged according to the demands of the environmental and social situations encountered. Differences between what individuals did and what they were supposed to do probably provided an important stimulus for change.

Some of these processes have been documented in studies of the development of Anglo-Saxon law. What had been folk peace became the King's peace, and offences earlier seen as offences against an individual or the community came to be seen also as offences against the King's peace. Authority for retribution, originally held by the wronged individual and his kin and later by the community, became vested in the King. Punishment often involved social ostracism and/or both retribution to those wronged and an element to the King for breaking the King's peace.[7]

The increasing complexity of human groups and individual ownership of property led to the formation of a hierarchical structuring, with leaders at every level:[8] medieval society with a king, barons with their subjects, knights with their retainers, and so on provide a familiar example. Within each subgroup the followers would have certain duties to the leader, such as the duty to fight for him in battle, and the power of the leader would be accepted because of certain services he performed in reciprocation, such as protection, or the granting of grazing or fishing rights. In other words, the incumbents of each role at each level would have certain rights and certain duties. The performance of those duties could be both a moral and a legal issue.

As societies became more complex, more complex legal systems would have been required. It would have become important that these should involve not only the laws, but also institutions for amending, applying, and enforcing them. That would have required a code of conduct for those who performed these tasks—an issue to which we shall return.

The direction of the further development of a legal system would depend on the circumstances of the society.[9] As discussed earlier, discrepancies between how people behave and how they are supposed to behave probably provide the most important force for change. The history and present circumstances of the group are important for public acceptance of innovations, their durability, and their nature. As just one example, the members of the early kibbutzim had come from totalitarian regimes, such as Nazi Germany and Stalin's Soviet Union, where rigid rules were enforced. As a reaction against this, it was first assumed that public order could be maintained by shared ideological understandings. However, it soon became apparent that formal regulations were necessary. Offenders came before the general assembly of the kibbutz. At that time the kibbutzim were struggling for survival under difficult conditions, so the punishments imposed on offenders were determined not only by the severity of the offence, but also by the value of the offender to the kibbutz as assessed by his past behaviour.

If these speculations about the evolution of legal systems are correct, and they can be supported by considerable, though largely circumstantial evidence, two points are relevant to the present context. First, legal systems emerged from, or in parallel with, shared understandings about acceptable behaviour, that is, from elementary moral systems. Many laws are clearly formalizations of moral precepts, forbidding actions to others that one would not like to have been directed by another to oneself. Second, they were often formulated and enforced by a king or ruler who may have acted partly or mostly from self-interest. As we saw in Chapter 2, this is also the case with some moral precepts.

SIMILARITIES, DIFFERENCES, AND CONFLICTS BETWEEN MORAL PRECEPTS AND THE LAW

GENERAL

The relation of law to morality is a matter that has been primarily of concern to legal theorists, but becomes of important general concern when attempts are made to change the law. What follows is not intended as a survey of the relations between publicly accepted morality and the law, but rather to highlight a few central issues, with emphasis on the extent to which the legal system reinforces or distorts moral precepts. Of course, legal systems differ over time and between states: references to 'the law' in what follows refer primarily to English law.

Many generalizations about the relations between law and morality are flawed by the assumption that each is homogeneous. They are not. Some laws are prescriptive, some proscriptive;

some serve the general good, some that of a minority. They may serve either to prevent or to maintain inequalities between individuals. Not all laws are (directly) coercive, some are permissive and some involve the distribution of resources. It will be apparent from Chapters 1 and 2 that moral precepts are similarly diverse both within and between societies. So generalizations about the relations between the law and morality are not to be trusted: while the comparisons that follow are on the whole valid, exceptions can be found to many of them.

DIRECTION OF EFFECTS

Does morality affect the law, or vice versa? Discussions of legal issues often use the terminology of morality: people talk of justice, fairness, rights, and so on. Arguments about the legality of abortion, or contraception, revolve around people's moral beliefs. This would seem to indicate that laws are based on morality, though it could be just legal-speak used to justify law, or to make it acceptable to the general public. By contrast, some legal authorities have sought to dissociate the law from morality, claiming that law in practice should avoid moral arguments, and that moral virtue is not necessarily a prerequisite for the validity of any law.[10] This may be justifiable on pragmatic grounds. But that view is argued on the basis that any other course would involve treating one case differently from others: in other words it would not be fair. That in itself is a moral argument. Again, in arguing that similar cases should be treated similarly, we can legitimately ask about the criteria by which similarity is assessed: this also is a moral matter. And cases arise when the law as written is not adequate: moral arguments may then be used, for instance taking into account the intention behind past judicial decisions even though the cases were not exactly comparable. In any case, discussions as to whether a law is 'good' or 'bad' are frequent. When we ask whether a law is moral, by what criteria can we make a judgement? To say that a law is or is not moral presupposes moral principles or precepts by which we assess the issue. One may conclude that laws are likely to be interpreted in moral terms. It is generally true to say that, in a democracy, laws that are not compatible with the morality of the society, or of its more influential members, would be unlikely to command acceptance in the long run.

But the influences between morals and law are not all one-way, for the law may influence publicly accepted morality. For example, once rationing had been legally imposed in the UK in World War II, illegal purchases of rationed goods were seen as immoral, presumably because the law was seen as consonant with fairness. Although much law is based on moral principles, there are times (like killing in war, Chapter 9) when law is seen as permitting or requiring behaviour normally seen as amoral or immoral.

SOURCES

Because most people see the law as man-made, but morality as either god-given, intuitive, or with origins deep in the past, we tend to ask whether the law is in keeping with morality, rather than vice versa. Moral precepts, as aspects of culture, are a product of a two-way relationship between what individuals do, or think about what they should do, and the norms and values prevailing in the culture. On the present view, as discussed above, legal systems originated from shared understandings or elementary moral systems, resulting from a formalization and elaboration of moral precepts (pp. 55–8). Thus both morality and law are ultimately social constructions, but constructions facilitated and constrained by basic human characteristics.

In particular, the principle of prosocial reciprocity provides a basis for most moral precepts concerned with personal relationships or relations within the community. Prosocial reciprocity and commitment similarly provide a basis for some aspects of the law concerned with relations between individuals, even in rather abstract contexts (e.g. contract law).[11] Reciprocity also contributes to the determination of punishments (see below), though group coherence may demand also an additional penalty (pp. 56–7).

So-called 'Natural Law' holds that there are objective moral principles that depend on the nature of the universe and can be discovered by reason. Although I have argued that morality is based on human nature, what is natural is not necessarily right—nor, for that matter, is what is seen as right necessarily natural. The discussion in the preceding chapters indicates that, ultimately, most basic moral principles are pan-cultural not because they are self-evident, nor because they accord with common practice, nor because they are divinely inspired, but because individuals do better in cohesive groups.

Nevertheless, because most law has its roots in moral principles that in turn are rooted in evolutionary principles, a biological viewpoint can provide some insights into the nature of our laws. For instance, genetic relationship is taken into account in deciding who should inherit the property of a person who dies without making a will, and rape is proscribed more rigidly than other forms of physical harm that do not have reproductive implications. Indeed the very concept of personal property has its origins in our prehistory: it may be related to the propensity, common to many species, to defend territories or resources.[12]

TEMPORAL CHANGES

Both morality and law are labile, though their effectiveness depends in part on their being perceived as absolute. What is seen as morally unacceptable behaviour at one time may be seen as quite ordinary later. Precepts seen as 'right' may later be deemed to have been improperly restrictive or immoral. Similarly, laws

may change with time: new laws are made and others abandoned or revised as the needs or nature of society change. The
mutual influences between what is seen as morally respectable
and legally permissible, and what people do, has already been
discussed in the case of divorce. Until recently (though no longer)
the changes in both law and morality have been so slow as to
make them barely perceptible to most people.

At any one time morality and law may not be completely in
harmony with each other. At times the law may be in advance
of morality: many would hold that this was the case with the
(relatively recent) permissiveness for homosexuality, which is still
seen as immoral by many within the UK and as illegal in many
other cultures. In other cases the law is seen by many (though
perhaps not yet the majority) as having lagged behind, as in its
recognition of single-sex partnerships.

As we have seen (pp. 12–13), changes in morality usually come
about through the dialectics between what people do and what
people are supposed to do. Sometimes they result from pronouncements by religious or secular authorities, but these are
seldom totally unrelated to public opinion or, if they are, they
are less likely to be viable. For example, at the time of writing
there are marked differences between papal pronouncements
condemning contraception and abortion as morally wrong, and
practice in many Catholic countries. Can we expect that, in due
time, papal pronouncements will change?

Except in circumstances deemed to involve a national emer-
gency, such as war or the threat of terrorism, the framing of
new laws is based on, or must take account of, publicly accepted
morality. The US Constitution recognizes certain fundamental
human rights, thereby formalizing what was previously generally
recognized or seen as desirable by its founders.[13] In practice some
changes in the law tend to be influenced more by the morality of
the elite than by that of the population as a whole. Thus in the

UK the legal status of homosexual acts was changed as the result primarily of the influence of more educated individuals.

The important issue here is that, although moral precepts and laws must be perceived as absolute, neither are immutable and the validity of both can be questioned.[14]

THE NATURE OF THE INDIVIDUAL

Criminal law, concerned ultimately with the well-being of the community, implies a distinction between law-abiding citizens and lawbreakers, and one of its aims is to reduce the impact of the latter on the former. It must necessarily focus, therefore, on the selfish assertiveness of individuals, and pay much less attention to the potential for prosociality present in all individuals. (This is less true for contract law, company law, and the law of partnership, which some would say have more emphasis on individual prosociality.)

Legally, convicted criminals are treated as non-persons who do not qualify for certain human rights, such as freedom of movement. The moral status of denying the right to freedom of movement to those who are merely suspected of terrorism is currently contested: the lawmakers must balance the safety of the general public against the general importance of human rights.

INTENTIONALITY

Morality and legal systems generally assume the existence of free will. Moral philosophers have discussed whether behaviour can be truly moral if not intentional. The view taken here is that moral behaviour may be automatic, stemming immediately from the principles and precepts in the self-system, or intentional, in the sense of being the result of conscious deliberation: all intermediates are, of course, possible. In some cases, the legal guilt of an offender depends straightforwardly on whether he broke the law: speeding offences, for instance, are such whether or not the driver knew or cared that he was exceeding the limit. But in other cases, the guilt of a defendant, or the punishment he receives, is influenced by whether the act required intention, knowledge of the results that would follow, recklessness, or mere negligence. For example, the crime of fraud requires that the defendant intended to deceive and defraud his victim. Inevitably intentionality, or its absence, is not easy to establish: in some legal systems evidence for an immediate response to provocation can affect the assessment of intentionality. Intentionality is therefore a salient characteristic in some but not all judgements of both legal guilt and moral transgression. However, morality, being less formalized than the law, can accommodate continua between intentional and unintentional more easily, and moral culpability can be diminished by the absence of intention. In law, by contrast, a clear distinction between guilty and not guilty is often necessary.

In this context it must be acknowledged that the nature and even existence of free will is coming under increasing scrutiny. Some neuroscientists maintain that the more that is known about brain mechanisms, the less scope there is for the concept of free will. For instance, a problem with the plea of insanity could arise from the fact that certain brain lesions can produce an individual capable of differentiating right from wrong, but nonetheless incapable of appropriately regulating his behaviour.[15]

THEORETICAL BASES

Legal theory often depends on contrasting principles, such as justice, equality, or utilitarianism (see p. 86). In the historical genesis of morality, whether the principles of equality, equity, need, or justice should be given priority in a just decision, though not unimportant, was a secondary consideration, probably depending on which was most conducive to group integrity in the conditions prevailing. In dividing resources, neither morality nor the law produce clear guidance on which of these guiding principles of fair distribution should be followed. Equality, equity, and need are most likely to satisfy those involved, but utilitarian distribution may favour the cohesion of the group and be in the interests of the majority.

FUNCTIONS

Moral precepts are concerned primarily with relations between individuals, though sometimes applied to relations between the individual and the community. In democracies, most law has been formulated in the interests of the society as a whole (or of the more influential individuals within it). As the example of the Israeli kibbutzim suggests (see above), in practice the smooth functioning of society requires a legal system as well as moral precepts to restrict the activities of free-riders.

Some laws may infringe the rights of the individual for the public good, as with the laws to deal with terrorist suspects enacted recently in the UK. In some cases, for instance in wartime, laws are imposed for the public good but against the wishes of the majority. Occasionally, laws are imposed against both the wishes and the good of the majority, as was the case during Apartheid in South Africa.

SCOPE

Morality and the law do not always cover the same ground. In societies with some separation between the sacred and the secular, the law alone would be insufficient to maintain the cohesion of society: moral prescriptions are essential.[16] Some actions that are generally considered as moral or immoral are outside the scope of the law. In general, the law is concerned with the more extreme examples of what one should not do, while morality emphasizes also everyday misedmeanours and what one should do. For instance, it is regarded as morally right to give to charity, but (in the UK) there is no law that one should. In this case, then, morality has regard for the common good, but the law at most encourages donations. In normal circumstances and over trivial issues, taking more than one's share is a moral but not a legal matter.

However, some laws involve proscriptions. Thus in many legal systems parents must look after their children, doctors and teachers must look after those in their care, and in some systems individuals must look after their domestic animals. In some, teachers must inform the authorities if a child shows evidence of having been ill-treated outside school.[17] To some extent, appearances may be deceptive here: some laws are framed as proscriptions because of the difficulty of specifying precisely what one should do.

Infringements of many laws, such as those concerned with conventions that facilitate the smooth running of society (e.g. minor traffic offences), are not always considered as moral matters. Some practices, now considered as immoral, such as slavery, were at first accepted by virtually all, then seen by some to be immoral, and only later, after much and heated argument, made illegal. It is noteworthy that, in this case, the argument reduced mainly to one between those who thought slavery wrong and/or stood to lose by it, and those who stood to profit from it.[18]

In a few cases, such as forced intercourse between marital partners, both morality and law would now condemn extreme examples of actions which in themselves would not normally be seen as wrong. That there may be moral limits to the extent to which the law should be obeyed by individuals[19] is exemplified by Kohlberg's classic dilemma of the man who steals drugs that will save the life of his wife but which he cannot afford to buy.

RELATIONS TO PUBLIC OPINION

As discussed in Chapter 1, moral precepts must almost by definition be accepted by a large proportion of the group to which they apply. (This does not mean that individuals may not feel morally impelled to act in a manner that contravenes generally accepted moral precepts: their decision may be ascribed to what they feel to be more basic moral principles.) Laws are seen as imposed by authority but, in a democracy, any law that is not generally accepted will not be enforced easily, and the drafting of new laws usually takes into account their acceptability[20] as well as their enforceability. That laws must usually be acceptable to the majority can be taken to imply that they are expressive of the public will but, conversely, in some cases, such as homosexuality and rationing in wartime, public attitudes may also be shaped by laws.

VALIDITY

On the view of morality advanced in Chapter 1, and the speculations about the origins of legal systems outlined above, it makes no sense to ask if either moral precepts or laws are 'true' or 'false'. Both are social constructions. The perceived validity of moral precepts depends ultimately on their acceptance by individuals in the community in question. Though individuals may hold idiosyncratic moral views, for most of the time no distinction can be made between the moral opinions of the collective and 'truth'. Change usually comes when required by societal change or when one or more individuals perceive existing practice to be at odds with the basic principle of prosocial reciprocity.

We have seen that moral precepts are usually such as to promote the good of individuals, and this usually results in the common good. In the case of many laws, however, it is the other way round. Some laws made to promote the common good can be to the detriment of some individuals. For instance a law against driving fast may, if applied to a doctor visiting a critically ill patient, be to the detriment of the latter. The Apartheid laws in South Africa were in the interests only of a minority.

Both moral precepts and laws can be against the common interest. Laws can be used for coercion, for protecting those in power and preserving their status. But so also can generally accepted moral precepts: it is not always in the common interest to behave humbly to one's social superiors, though the Christian emphasis on humility may make it seem so. And precepts that make it defiling to have contact with a member of an 'inferior' group or caste, seen as moral precepts by the in-group, contravene the Golden Rule and are immoral as seen by outsiders. Similarly ostracism of an individual who has offended a social norm may be seen as correct by in-group members but immoral by outsiders.

MORALS AND CONVENTIONS

Some rules of conduct and some laws are concerned primarily with ensuring the smooth running of society, rather than with what is seen as 'right'. Some psychologists argue for a distinction between morals and conventions, arguing that they depend on different conceptual domains.[21] Morality is justified by the extent to which it promotes welfare, justice, rights, truth, and so on, whereas conventions involve merely understanding the society's rules. In the same way, some laws have a close relation to moral principles (e.g. laws relating to homicide) and others merely reify conventions (driving on the left/right side of the road). But these distinctions are far from clear, and most laws and moral precepts are such that in the end they promote the smooth running of the community.

ENFORCEMENT

The effectiveness of public morality is maintained by processes such as felt responsibility, guilt, and shame in the individual, by social disapproval, and by fear of punishment in this life or the next. Criminal laws stipulate how (within limits) offenders should be punished (see below).

CONFLICTS BETWEEN INDIVIDUAL AND GROUP[22]

Just like morality, law has to cope with goals that often conflict, most often the autonomy of individuals or the views of a group within the larger society on the one hand, and the common good on the other. The law, like morality, sometimes has to restrict a human right for the sake of the community. For instance, a law against racial abuse may restrict the right to freedom of expression, and laws designed to protect the community against terrorism may involve telephone tapping and other invasions of privacy as well as restricting the freedom of some individuals. While human rights must be considered as absolute, there are exceptional circumstances when they are withdrawn. Thus the right of free speech may be withdrawn from those who preach sedition. We shall discuss later other cases where rules must be seen as absolute, yet must have some flexibility.

There is always the question of how far laws that restrict the rights of individuals should be countenanced. In the UK homosexuality has been a focus of dissension. There were formerly deep differences over whether homosexual acts in private should be permitted. It is noteworthy that it is since such acts were deemed not to be a matter of public concern that there has also been a relaxation of feeling against them.

Sometimes restriction of rights is in the interests of the individual. A classic example is to be seen in the 'tragedy of the commons'. A common resource, such as a fish population, may be self-sustaining so long as individuals are forced to restrict their exploitation: if individuals take more than their allotted share, the resource may disappear. Another example is a law involving conscription for the armed services. In times of war it may be seen as serving the common good to require all individuals to serve the government's purposes: this may infringe the rights of individuals in the short term, but benefits most individuals in the community. A more everyday example of the law restricting individuals for their own good even when that is against their own wishes are some of the health and safety laws that make jobs more difficult to carry out.

Many laws protect the rights of the individual. On the other hand, some rights, even those that are more or less legally recognized, may be to the detriment of the common good: the right to bear arms in the USA, or the early twentieth-century laws denying women the right to vote in the UK, are clear cases in point. Some laws are to the benefit of both individual and community, like those requiring safety standards for cars.

But over some issues, feelings still run high. Many laws that are for the common good restrict individual autonomy. A law compelling motor cyclists to wear crash helmets was seen initially as unjustifiably restrictive by many. One that restricts cigarette smoking in public places is for the common good, and might help the smoker to give up the habit, but is felt by some as a gross infringement of the individual's right to freedom of action. Another example concerns the use of drugs. For many, drug use provides escape and relaxation. A few are harmed by their use. Banning of drugs results in illegal trafficking, with all the exploitation and corruption to which that leads. The result has been not infrequent disputes over precisely which drugs should be banned and resulting changes in the law. Usually the argument has focused on the harm to the drug-user, and much less is said about the consequences on the community. This issue is also much influenced by public biases, for tobacco and alcohol are virtually unrestricted, and there are few restrictions on the sale of unhealthy foods.

Finally, the moral consequences of some moral precepts or laws may change with circumstances. In discussing exchange theories of interpersonal relations we shall see the importance of returning obligations and keeping promises. But a promise to, or contract with, a person now deceased may give rise to legal difficulties and be detrimental to the good of others if the beneficiary dies: the problem then becomes a moral one.

CULTURAL DIFFERENCES

Some examples of the many problems raised by cultural differences in moral precepts and values were mentioned earlier. The laws of one society may be judged as immoral by outsiders. I have argued that one can properly judge aspects of other *cultures* that contravene the Golden Rule as unacceptable, but that that is a different matter from judging *individuals* who cannot have known otherwise because they were brought up in that culture. Similar problems occur within a society, and recently it has been asked whether immigrants from another culture should be made to conform to the customs and values of the host country by law. This can involve issues seen as moral by the immigrant, such as the wearing of a headscarf.

LAW ENFORCEMENT

We noted earlier that the formalization of a legal system required an apparatus for enforcing the law, and any such apparatus would require a code of conduct for those who operated it. That considerations of self-interest should be ruled out goes without question but, from the perspective of the present discussion, it is of great significance that such a code may require individuals to act in a fashion contrary to the everyday morality he or she would accept in other contexts. Thus in the British system barristers may be appointed to argue the prisoner's guilt or innocence, whether or not they know the true facts of the case and whatever they believe about the guilt or innocence of the accused.[23] The merit of this system is that it strives to ensure that the case is argued by those fully familiar with the law, and thus that the accused has a fair trial. For a barrister not to press his/her case to the full would be regarded as ethically incorrect because it would involve a dereliction of his duty as a barrister in the legal system. In other words, law enforcement has led to a system in which, in the interests of fairness, individuals may be required to act in a way contrary to the moral judgement they would make outside the court. A similar principle applies to judges: an especially acute problem arises from the requirement for British judges to hear appeals in some West Indian countries where the death penalty is still legal. A British judge who knows a man to be guilty may be required to condemn the prisoner to death against his own moral convictions concerning the death penalty and against the law of his own country.

PUNISHMENT

Here I am concerned primarily with criminal law. The criminal law differs from morality in that in general it is enforced by state-imposed punishment, while most infringements of moral rules that are not also legal offences are enforced at most by feelings of guilt, social rejection, or denigration. Laws are enforced by physical sanctions, morals by subjective ones—though, as we have seen, social disapprobation may be a practical handicap.

But the question of legal punishment poses significant practical and moral problems. In the first place, does it assume that the offender knew what he was doing and could have behaved otherwise? If the concept of free will should become whittled away by advances in neuroscience (see above),[24] the aim of punishment may come to be increasingly focused on its consequences for the offender and for society.

Again, precisely how can depriving an individual of liberty or even life be morally justified, for it is taking away the prisoner's rights and treating him/her as a non-person? Should the aim be to make the punishment fit the crime, to protect the community, or to prevent recidivism, or what? I consider here some of the moral issues that arise, but not with any implication that enlightened judges do not act in full awareness of their complexity. Judges have considerable flexibility in the penalties that they can impose, and must balance a number of principles, all of which involve ethical considerations.[25]

REVENGE

As we have seen, originally revenge, or threat of revenge, probably served as a means of maintaining peaceful relations within groups that lacked a legal system. The desire for revenge or retribution is probably a pan-cultural characteristic of humans, perhaps as the inverse of the principle of prosocial reciprocity. But we have seen that revenge can lead to escalation, and must be controlled. We have seen also that, in the development of Anglo-Saxon law, revenge came to be controlled by a generally accepted authority, and that retribution involved both a penalty for disturbing the peace (or the costs of legal proceedings) as well as retribution to the wronged party.

Some legal theorists see giving the offender 'just deserts' as a proper aim in itself, though supporters of this view may see punishment also as censuring the offender and as a disincentive for future offending. 'Just deserts' implies that sentencing should be related to the severity of the offence, and thus in turn the presumption that it is possible to rank crimes in order of seriousness. This could be an impossible task if diverse types of crime were being considered. In addition, retribution should take into account the harm done and degree of culpability, and also, perhaps, previous convictions. Culpability would require allowance in sentencing for the personality and circumstances of the offender. Even if done in all cases, this would reduce the chances of the punishment being perceived as fair by the offender, by other offenders, or by victims. Because offenders compare themselves with each other, and because victims may not be satisfied until those who have harmed them receive punishment seen to be just, fairness is important. Clearly there is room for considerable legal disagreement.

DETERRENCE

Some legal authorities give a high priority to deterrence as a principle of sentencing. The aim is to punish the offender sufficiently severely to prevent recidivism, and also, by example, to deter other citizens from committing similar crimes. The former goal implies that the penalty inflicted should be perceived as onerous by the particular offender. For that to be the case, his character, circumstances, and previous record should be taken into account. In addition, the offender's perception of the likelihood of being apprehended if he offends again may affect the probability of recidivism, and is also relevant. But taking such individual characteristics into account may detract from the perception of fairness by others. If deterrence of other possible offenders is a goal, care must be taken to ensure that the offender is not punished excessively, that is, more than the crime merits and thus unfairly, merely to ensure that it will have an effect on the community.

If recidivism is to be prevented, one aim of punishment should be rehabilitation. Rehabilitation usually involves support, counselling, and education or job training. It is probably most likely to be effective if the offender stays in the community, provided it is a tolerant community in which criminal values do not predominate. Some data on recidivism do not encourage optimism about the effects of attempts at rehabilitation, but much probably depends on the nature of the rehabilitation programme and the conditions in the prison. It can certainly be successful for certain types of individual. A rehabilitative technique applied indiscriminately across the board, regardless of the recipient, is unlikely to be worthwhile.

PROTECTION OF THE COMMUNITY

This is clearly a primary aim in the sentencing of criminals considered to be a danger to the public, and in the controversial practice of detaining those suspected of being potential terrorists. Whether the well-being of the community in itself provides an adequate justification for the use of imprisonment is likely to be a difficult decision. A period of incarceration or house arrest ensures that offenders are not able to offend again so long as they are inside. But if the community's welfare were the prime consideration in deciding the length of time for which the offender should be removed, it would be necessary to take into account the probability of his offending again, as indicated by his past history, his circumstances, and the extent to which he is judged to be a danger to the public. All of these are notoriously difficult to assess, so that penalties calculated on this basis are likely to be unfair to the individual.

If the protection of the community were the only justification for prison sentences, then fair retribution would not be an issue, and there need be no pretence that the punishment should fit the crime. Logically, the severity of the punishment might then be adjusted according to the difficulty of eradicating the offence in the community. But that is by no means always possible: drunkenness might be so difficult to deal with that, for the sake of protecting the community, it might merit life imprisonment. There are, of course, other issues, one being that the machinery of punishment is costly to the community, so the social benefits must not be outweighed by the costs.

If the well-being of the community were the primary aim, the effects of incarceration on the prisoner after release must also be considered. Prisons are never perfect and are often overcrowded: as a result imprisonment can be a dehumanizing experience. It is all too easy for warders to treat prisoners as a race apart, taking away their human dignity: this may have devastating effects when they are released. Association with hardened offenders may infect first-timers. Recidivism after a prison sentence is high, so punishments as administered at present do not seem very effective as deterrents and, apart from the period of imprisonment, have little effect on the community's security. In the long run, rehabilitation must be in the interests of society, but degradation and dehumanization of the prisoner make rehabilitation a less likely outcome. For certain types of offender probation orders may be more suitable. In the UK, Anti-Social Behaviour Orders can be issued preventing the individual from entering a certain area or associating with particular individuals for a stated period of up to two years. Disobeying the order becomes a criminal offence.[26] However, their effectiveness is debatable.

PARDON

On rare occasions, the offences of an offender or a category of offenders may be pardoned. This is usually a device calculated to enhance the prestige of the pardoner.

GENERAL

No one of these principles is perfect and ubiquitously applicable. Apart from the issues already mentioned, such as the difficulty of assessing 'severity' and the probability of recidivism, the principle of fairness demands that similar offences merit similar punishments. But that leaves out the question of degree of responsibility, including the mental state of the offender and the nature of his/her motivation. Special allowance may be made for the *crime passionnel*. Many would say that in theory justice should be emotion-free and take into account only the offender, not his relationships. But offenders offend for very diverse reasons, and sentences similar in terms of months of incarceration may be experienced quite differently by different offenders according to how it affects their own self-image and their relationships with others. Thus a prison sentence may have a devastating effect on family relationships: this is especially the case with young offenders, for whom a period in an institution may sever family ties.

Thus the rights of the individual and the good of the community are inextricably entangled. If the purpose of the law is to protect the community, it also has a duty to protect the individuals who constitute the community, and the prisoner is still part of the community—or will soon be restored to it. For that reason alone, rehabilitation should be a high priority.

Such considerations suggest that a hierarchy of principles should be applied, and this has been attempted in some legal systems, but the question still arises, should the hierarchical order be the same across all offenders and all offences? In any case, for both moral and pragmatic reasons, sentences should be as reasonably lenient as possible—morally because any punishment, and especially incarceration, involves some measure of removal of the offender's personhood, and pragmatically because the costs to the state, and thus to other citizens, of prisons and related services are high. For the latter reason, and for their probably greater rehabilitation potential, non-custodial sentences are to be seen as preferable where possible. These must, of course, ensure that the offender is not seen to pose a threat to the community.

PUNISHMENT AND THE VICTIM

Also, what about the victim? Retribution may inflict on the offender injury comparable with that which he has caused, but that does little good to the victim, except subjectively. For many offences compensation of the victim is impracticable, but in some it can be appropriate, especially in some cases of corporate misdemeanour (see Chapter 8). That the victim's feelings are not unimportant, however, is shown by their apparent satisfaction after offenders have been successfully prosecuted and sentenced. Even when possible, material compensation may not be what is most important to the victim; acknowledgement of guilt and/or apology may be more satisfying. In some cases it can be helpful for the victim and offender to meet and discuss the crime, for this can enhance understanding of what has happened and why.[27]

THE PROBLEM OF THE ELITE

A further issue relevant to punishment, though of fundamental importance, can be mentioned only briefly. Legal systems, as we have seen, are derived from moral systems that deal primarily with relationships between individuals. Ideally all offenders should be treated in a similar way, but are they? Many crimes against individuals are committed by states, corporations, industrial enterprises and suchlike. Those held to be responsible for such crimes tend to be amongst the elite members of society, those who suffer ordinary citizens.

I mentioned earlier that change in the laws about homosexuality in the UK largely reflected the views of the more educated individuals, and might well not have received majority support in a referendum. Do other laws unfairly represent the views or, worse, the interests, of particular sections of society? Box[28] has argued provocatively that laws may favour the upper classes. For instance the intentional killing of one individual by another is described as murder, but predictable deaths resulting from corporate failures receive lesser designations. Examples are employers' failure to maintain proper safety standards in factories, the aggressive marketing of a substance whose full effects were not yet known, or car manufacturers failing to recall vehicles known to be defective. A well-known case of the latter concerned a motor company's knowledge that the design of the fuel tank in one of its models was defective and fractured easily, especially in rear-end collisions. It was alleged that they decided to proceed with the design because the cost of remedying the matter would be greater than the predicted settlements for damages that would result. It has been estimated that more than five hundred deaths from burns resulted. Another case is the drug thalidomide: around eight thousand mothers who had taken the drug produced deformed babies. And the relations between asbestos and lung cancer were apparently well known to at least some of the manufacturers from their own research, but suppressed. Many other examples could be cited. It is worth noting that part of the difficulty lies in attributing moral characteristics or mental states to corporations and associations. Most laws depend on establishing a particular moral responsibility (*mens rea*) and thus a 'directing mind and will'. This makes prosecuting a corporate body difficult.

FORMULATING NEW LAWS

Given that morality has bases in human nature, and most laws are related to morality, can evolutionary theory provide any guidance for lawmakers? The answer is yes, but rarely: for instance, the law that parents can be prosecuted for neglecting their children is compatible with biological principles. But evolutionary considerations must be used with care, for laws may have unforeseen consequences. Take for instance the fact that a high proportion of incidents of infanticide are performed by step-parents.[29] This could lead child protection agencies to look especially carefully at families containing a step-parent, but they would then run into the danger of unfairly stigmatizing many individual step-parents. Laws against abortion may lead to an increase in parental infanticide by parents of unwanted children and by rape victims.[30] Evolutionary considerations also suggest that legal constraints on basic human propensities, and especially those that concern sexual behaviour and prostitution, will be especially difficult to enforce.

CONCLUSION

Piecing together evidence from diverse sources, it is possible to suggest a speculative scenario for the cultural evolution of both moral and legal systems. Modern hunter-gatherer groups tend to be egalitarian, with acceptable behaviour defined by mutual understandings. Individual assertiveness is kept in check by social pressure, sometimes involving communal force, or revenge by injured individuals or their kin. As leaders emerged, they acquired the right to impose punishment for offences: punishments often involved both retribution and an element for breaking the peace. While morality emerged from shared understandings about what behaviour was and was not acceptable, legal systems subsequently followed their own courses and came to differ from morality in many ways. Where similar issues are concerned, the law is concerned with more extreme departures from societal rules than is morality. Both moral precepts and laws can change with time and differ between groups or societies.

Legal systems discourage individuals from disregarding moral precepts. Like morality, the law both influences and is influenced by how people behave. However, a legal system takes on a life of its own, so that morality and the law do not precisely coincide. Discrepancies have arisen because, while morality is concerned with a wide range of infringements, the law is limited (in most cases) to the more extreme ones. In addition most moral precepts originally operated in two-person interactions and relationships (though taking account of the well-being of the community), while most laws are concerned primarily (though not exclusively) with the well-being of the community. Law is thus less concerned with consequences for the relationships between individuals than are moral precepts. Where laws are concerned with interpersonal relationships, it is usually with half an eye on the good of the society as a whole. Laws designed for the good of the community may override individual rights, though some laws recently enacted defend the individual from excessive public authority. With some exceptions, such as some laws dealing with human rights, laws have been formalized more rigidly than have moral precepts. In addition, again with some exceptions, most laws are proscriptive, while morality is concerned with what one should do as well as what one should not.

Perhaps a non-legal outsider may be allowed to point to an anomaly relevant to the theme of this book. While the origins of legal systems are to be found in moral rules, their operation may require moral rules to be broken. The case of barristers who must prosecute or defend individuals with a greater certainty than they entertain themselves has already been mentioned. But, as seen from the outside, the administration of a legal system from judge to warder entails doing to others what one would not wish to have done to oneself. Yet the purpose of the whole system is to maintain moral behaviour in the community. We need and approve of the legal system and the work of all those involved, and our approval enables them, if they consider the matter at all, to feel that they are properly doing their duty to the society. We shall return to this theme in later chapters.

4

Exchange and Reciprocity
Conflict in Personal Relationships

INTRODUCTION

Personal relationships provided the context in which moral precepts emerged, for the integrity of a small group depends on the relationships between its members. This chapter considers some aspects of how personal relationships work, and has two sections. The first introduces perhaps the most important factor in the day-to-day conduct of personal relationships—reciprocal exchange. Related to that, many human values ensure that personal relationships run smoothly, and have presumably been selected, biologically or culturally, to that end.

But relationships do not always run smoothly. The second section outlines some of the problems that arise between individuals in close relationships. While the moral precepts by which we try to lead our lives were elaborated over evolutionary and historical time to smooth the course of one-to-one relationships, conflicts must always have occurred, and social life has become more complicated than in the past. In addition, precepts originally concerned with one-to-one relationships are applied also to how individuals should behave to groups or categories of individuals. It is therefore necessary to mention also some difficulties that occur in the relations between individuals and groups or group values. In both cases, problems arise largely because the parties involved have different perceptions of what is 'fair' or what would constitute 'correct behaviour'. In that case, how can conflicts ever be resolved?

EXCHANGE THEORIES

Whereas previous chapters discussed the forces that shaped the nature of morality in the past, this one is concerned with factors operating in the here and now. A theoretical approach, or rather a family of approaches, that has been fruitful in understanding the dynamics of interpersonal relationships, comprises the several varieties of exchange theory.[1]

Exchange theories have in common the view that relationships between individuals involve processes of exchange, each partner incurring costs in the hope of receiving future rewards. In brief, A helps B in the expectation that B will help A reciprocally in due course. Proto-reciprocity starts with very young children: games involving throwing a ball to and fro involve trust that the partner will reciprocate.[2] It will be apparent that exchange theories are entirely compatible with a principle of prosocial reciprocity, and we shall see that this principle has provided the basis for many human virtues, though the emphases differ somewhat between cultures.

Exchange theories were developed initially for the context of employer/employee relationships, but have been extended to close personal relationships. To some the suggestion that personal relationships could have anything in common with transactions in the marketplace will seem deeply shocking, but we shall see (Chapter 8) that the morality of close personal relationships is basic to, but has been distorted in, the morality of the marketplace. And once again, we shall see that much stems from the interplay between prosociality and selfish assertiveness.

Several aspects of exchange have special importance in the present context. First, exchange as it is seen in human relationships would not be possible without mutual understanding and also some sense of property.[3] The latter may have originated with physical possession, a challenge to ownership leading to defensive aggression, but presumably over time possession has acquired a more abstract connotation.

Second, if A helps B in the expectation of future reciprocation, A must trust B not to abscond, for A would then lose his expected recompense. This usually means that he must perceive B as committed and trustworthy—and vice versa. Commitment and trust are indeed essential for many aspects of social life. A is more likely to trust B if B is a relative or member of A's group, for B is then more likely to share A's values and norms than a stranger: this may lead to the development of group markers (see pp. 47–8).[4]

Third, a relationship involves a series of interactions over time, and the course of each interaction may be influenced both by the outcome of previous interactions and by expectations for interactions in the future. Therefore, in the long term, continuity of interactions depends on both partners to the relationship being satisfied.[5] For A to maximize her outcomes in a relationship with B she must consider not only the rewards and costs to herself that result from her actions, but also those of B. If she does not, B may opt out of the relationship and A will have lost any future rewards. Furthermore B will not remain in the relationship, and thus will not provide the expected rewards, unless A's initial behaviour was prosocial. A negative approach to another individual is likely to provoke a negative response, and interaction may cease. In other words, for successful relationships, reciprocity and prosociality must be linked.

Fourth, satisfaction implies that each party to the exchange sees it to be fair. Thus A must perceive that the costs she has incurred in acting prosocially to B are balanced by the rewards that she receives, or expects to receive, from B. This can give rise to a problem, for perceptions of fairness, or in other words the perceived rules of exchange, differ between individuals and with the situation. There are three basic criteria applicable both to personal relationships and groups.[6]

(i) Equality: everyone deserves equal outcomes.

(ii) Equity: everyone's outcomes should be related to what he has put into the endeavour. This may include both the costs (labour, etc.) that he has incurred and his 'investments', in the sense of what he is invested with—skills, expertise, social status, and so on.

(iii) Social justice: everyone deserves outcomes in accordance with his needs.

A further type of justice, 'legitimate competition' will be mentioned in the context of business relationships (Chapter 8). Sharing according to the principles of equity and social justice demands an understanding of how others feel, and thus requires considerable cognitive development before they appear.

In virtually all areas of human life, 'fairness' is a crucial issue. But what matters for a smooth relationship is that each should receive what he *perceives* to be fair. There is therefore plenty of scope for disagreement: an exchange perceived as fair by one partner may not be so perceived by the other both because each will have his own biases and because they may use different rules of fairness.

Interestingly, not only those who feel under-benefited, but also those who feel over-benefited, may feel dissatisfied with the relationship. For instance, in a study of dating relationships, the relationships of those who felt themselves to be equitably treated were the most stable. Those who felt themselves to be over-benefited felt about as much satisfaction, and more guilt, than those who felt themselves to be under-benefited. In another study, both women who felt deprived and women who felt unfairly advantaged in their marriages showed greater desire for extramarital sex than those who felt themselves to be in an equitable relationship.[7] The generality of such studies needs to be assessed and other interpretations eliminated, but they provide strong support for the view that people are guided, perhaps unconsciously, by a rule demanding fair reciprocity or justice in relationships—in other words, by a social contract. Our propensity to help others less fortunate than ourselves may be related to this.

Of special interest in this context is evidence from experimental games that people's propensity to be fair in exchanges may override their rational self-interest. The game involves two players, one of whom (say A) is given a sum of money to distribute between the two of them. The conditions are that if B accepts what A gives him, each can keep the money he has, but if B rejects A's offer, neither player gets anything. It might be expected that A would offer B a small percentage of the total, and that B would accept, for if B did not accept, he would not get even the small amount offered. In practice, in Western societies A offers on average about 40 per cent, claiming that he should merely have a little more than B because he has the money. Offers of less than 30 per cent are frequently rejected by B.[8] It therefore seems that humans have a propensity to be fair. Of course, one must always consider the possibility that individuals act prosocially because they are responding to subtle cues,[9] or that they have grown up to believe that they are always being observed by others (real or imaginary): studies of the development of fairness are far from complete. In any case, the proportions offered or refused in these games differ between societies in ways that often seem to be related to the ecological conditions and/or conventions in the society in question.[10] For instance, in a New Guinea group in which receiving a gift involves a strict requirement for reciprocation, the rate of rejection was high, but in a group with low norms of cooperation rejection rates were low.

Fifth, exchange is not all there is to close personal relationships, but it forms the basis of many of our values and the precepts that guide our behaviour. One would not want to enter a relationship with someone whom one perceived not to be honest, as he might cheat. One must trust the partner in an exchange, and one's ability to trust depends in part on previous experience leading to the internalization of norms of trustworthiness and one's ability to detect trustworthiness in others. As mentioned previously, there is in fact growing evidence that we are rather good at detecting the infringement of moral codes,[11] at detecting dissembling,[12] and at remembering those previously labelled as cheats.[13] Certain emotions, such as righteous indignation, guilt, and shame, help to ensure that exchanges are satisfactory to both parties. This does not mean that such qualities are consciously related to exchange by the participants in relationships: they have been reified in the course of moral development as qualities important as guides to action. Virtues such as honesty and trustworthiness are in no way belittled by the recognition that their value stems ultimately from the selfish assertiveness that demands at least a fair deal in exchanges. Natural and cultural selection have given rise to the most revered human values.

In addition, we prefer to interact with individuals whom we perceive as sensitive and having skill in interpersonal relationships. This embraces the extents to which A sees B as B sees B (i.e A understands B) and to which B feels that A sees B as B sees B (B feels that A understands her), for these also can be seen as predictors of successful exchange. Understanding the other's point of view can also provide a route to the solution of conflicts.

Honest and fair dealing is also supported by the existence of sanctions. The latter may lie merely in the partner knowing that you will not interact with him again if he reneges on the present agreement, or on the possibility of revenge, or on social sanctions imposed by a third party.[14] Third party intervention can be seen as involving a sense of collective responsibility to prevent cheating. A third party who sees another being unfairly treated feels morally outraged and may intervene. Indeed it is held to be morally irresponsible not to punish others who are seen not to be living up to the accepted standards (see pp. 38–9). Thus the morality of interpersonal relationships involves the social group or community: this is in keeping with the view that prosociality has been maintained through the advantages of living in a harmonious group.

A sixth issue important in the present context is that the costs incurred and rewards received may not be measured in the same objective currency. If A buys a material item from B, he may be repaid with money. One aspect of this is that exchange can occur between individuals who are not social equivalents: an employee gives services in the expectation that the employer will give him a monetary reward. That costs and reward are not measured in the same currency makes deciding what is fair more difficult: a reward may consist of honour, public recognition, love, or other intangibles. An interesting case arises in the settlement of some legal cases. When a defendant has left money on the table in settlement of a claim, the plaintiff may leave it there if there has been a heartfelt apology from the defendant: what the plaintiff required was primarily a restoration of dignity through vindication of the claim.[15]

Furthermore, the reward may be delayed. The lover gives his lady a bouquet, hoping not for flowers in return, but reciprocation of another kind at a much later date. In the meanwhile the donor may be satisfied by an expression of gratitude. The gratitude acts as a sort of IOU until reciprocation can occur.

But gratitude has come to mean more than that. People behave prosocially to others whom they never expect to see again—for instance by giving up a seat in a train, or returning a dropped purse to the police station. In such cases the gratitude of the recipient acts as sufficient recompense. Acknowledgement for behaving prosocially confirms the values that guided one's behaviour and thus enhances one's self-esteem—one feels a better person, because one has lived up to one's own standards. The recipient's gratitude is even more acceptable if it is given publicly, for if it is witnessed by others, it marks the individual as prosocial, and others are more likely to deal with a person who has a reputation for prosociality.[16]

There is, of course, no implication that exchange is all there is to personal relationships. But it is noteworthy how many aspects of personal relationships seem to result from their foundations in exchange. Exchange theories offer generalizations about the ways in which people behave in personal relationships, and they add strength to the view, advanced in Chapters 1 and 2, that the principle of prosocial reciprocity is basic to personal relationships.

THE UBIQUITY OF CONFLICT

If moral principles and precepts were absolutes and never incompatible with each other, there would be no difficulty in distinguishing right from wrong. But moral precepts often turn out to be inadequate, inapplicable, or inappropriate in particular situations. Neither the claim that what is self-evidently good, nor the claim that what is in most people's interests, should always be chosen is ubiquitously applicable. No doubt this is exacerbated by the complexity of modern life, for individuals are often faced with incompatible oughts. In many cases, the conflict is resolved unconsciously, or at least relatively easily, but in other cases decision-making may be a long and agonizing process. It may lead to moral satisfaction if the individual is sure he has taken the right course, but to guilt or shame, requiring apologies or atonement, if the perceived actions are incompatible with internalized moral precepts.

In the rest of this chapter I consider some causes of conflict in personal relationships, and of conflict between individuals and the well-being of the group, that go beyond merely pragmatic issues. You may find it of interest, as you read, to imagine a situation in which you were involved, and consider how you would act, and how you would judge the person or persons with whom you were in conflict. The issues here are diverse but fall roughly into the following, admittedly overlapping, types:

Conflicts Arising from the Nature of Relationships

DIFFERENCES BETWEEN WHAT INDIVIDUALS PERCEIVE AS FAIR

Reciprocity requires that exchanges be perceived as 'fair', but the participants in a relationship may have different perspectives. As noted above, each participant should see not only his own outcomes as fair, but also those of the other individual involved, and should also perceive that the other sees the exchange as fair. Even the participants in a two-person relationship may differ in the criteria that they apply. For instance, a teenage daughter seeking an advance on the allowance that she gets from her parents might base her case on needs, while the parent might ask what she had contributed to the family, favouring the criterion of equity. Again, should relations between the sexes be governed by equality, or by their differing needs? In financial transactions the rewards and costs for each party are usually fairly clearly defined, but if A does a kind turn to B he thereby creates a diffuse obligation for B to reciprocate without specifying what would be a fair exchange.

In games involving several players, V. L. Smith[17] claims that when there is no means to differentiate individual contributions, people favour equality; but if contributions can be differentiated, rewards tend to be given in proportion to the contributions the individual has made to the group.

Further complications arise from the fact that the two parties often use different currencies. When one individual performs a service for another, say a doctor for a patient or a lawyer for a client, reciprocation is likely to be in a different currency. Should the aim be to see that each gains equally from the transaction, even though that means weighing chalk against cheese? Should one take into account the time necessary for doctor or lawyer to acquire the necessary skills, or the need of the patient or client, or her financial resources? Surely the decision should not be according to the law of the marketplace, the doctor taking as much as he can get. And should the yardstick be similar for different exchange relationships—say nurses and engine drivers? Such issues are usually influenced by politicians, who consider the needs of society as a whole (e.g. there are too few nurses), or by other high status members of society who establish conventions, often to their own advantage. In other cases there may be no agreement as to what is fair. As we have seen (pp. 32–3), while parents are morally required to look after their children, a child may expect more than the parent is prepared to give. The parents may then try to persuade the child of the correctness of behaviour that is in reality biased towards the parents' own interests, but the child may be hard to convince.

A related issue is what individuals consider they deserve. Military officers expect to earn more than lower ranks, and executives expect to earn more than those who work at the bench.

Individuals may feel that what they deserve is augmented by strength, beauty, wealth, social status, sex, and so on, according to the particular values emphasized in their culture. Older children may expect and get larger presents, or more pocket money, than younger ones, who then perceive themselves to be unfairly treated. Alternatively, a handicap may be seen as deserving special treatment. Formalization of such conventions by those in power may make it easier, or be designed to make it easier, for those less well endowed to accept their lot, but disputes often arise.

INCOMPATIBILITY BETWEEN RIGHTS CLAIMED OR PRECEPTS HELD

Many conflicts involve the differing requirements of apparently incompatible moral precepts or values. As we have seen, some family disputes involve conflict between what parents see as proper care and what children see as their right to autonomy or privacy. Aged parents may see it as their right to be looked after by a child, who may feel that this conflicts with her right to education, or to realizing her talents to the full.

Slightly different issues are involved when behaviour perceived as morally correct by the actor conflicts with the wishes of the recipient. Outside the family, giving help to the sick or aged may be perceived by the latter as damaging to their self-esteem. When a young person altruistically offers her seat on a crowded train to an older person, the latter may feel that he or she really must be getting old.

INCOMPATIBLE PERSONAL CONCERNS

Relationships involve responsibilities: friendship involves the obligation to help the friend in need, parenthood the obligation to look after one's children, and so on. The most obvious and commonest cause of conflict occurs when personal needs or desires conflict with the responsibilities inherent to the relationship. My friend is ill and needs me, but I have to finish this job today.

A more interesting case is where the distinction between honesty and dissembling is hard to draw. It is good to be honest and direct in one's dealings, but there are contexts in which honesty has to be tempered with discretion. Too great an insistence on directness in communication, as well as too little, can be devastating for social relationships: to take a hackneyed example, it may not be wise to say everything that one thinks about one's mother-in-law.

Conflicts Arising from Societal Complexity

The morality of interpersonal relationships originated in relatively simple societies, and was concerned primarily with interactions and relationships between two individuals. Most of the issues that follow stem from the sheer complexity of modern societies or from differences in moral codes within and between societies.

MULTIPLICITY OF CONSEQUENCES

Every action has many consequences, and it may well happen that some are seen as morally good and others as bad. The consequences may involve not only the other party in the interaction or relationship, but also others in the community. For instance, divorce has consequences for the parties involved, for the children of the marriage, and for other family members and friends: it may raise religious issues, and every divorce may affect societal norms. The outsider can take one of two views. One is to take a stand on a supposed absolute, such as 'Divorce is wrong'. This might also be perceived as benefiting society as a whole through the maintenance of values. Or a course could be chosen that seems to involve the least harm for all concerned—partner, self, children, wider family, and so on. In practice, when those actually involved take the latter view, they are likely to bring their own biases to their decisions, biases that would depend largely on compatibility with the values incorporated in their self-systems.

PERSONAL AUTONOMY AND RIGHTS

The rights that an individual is seen to have emerge gradually in development as the result of interaction between the individual's self-assertiveness and the demands of caregivers or society. They concern actions that the individual considers outside justifiable social regulation by others. We have already seen an example of an emerging sense of rights: adolescents in our society may come to feel that they have the right to determine the tidiness of their bedrooms, while parents may feel that their own well-being or that of the family requires that they have some control. For adults, certain rights are formalized in each culture (e.g. freedom of speech, assembly, movement, and so on), and may be regarded as absolute. Although cultures differ in the rights that they recognize, and in the range of individuals to which they are granted, they mostly follow from the principle of prosocial reciprocity (see p. 85) and the Golden Rule, and are basic to democracy. Thus, by the Golden Rule, if I claim the right to my own opinion, I must allow others the right to theirs. Freedom of movement, and in particular the right to leave the group, have been emphasized by some writers as placing a limit on the behaviour of dominant individuals: if bosses become too bossy, group members may leave.

Individual freedoms, institutionalized in the USA as 'inalienable', can be seen as joint products of the Golden Rule and selfish assertiveness leading to personal autonomy. But what one wants for oneself is not necessarily good for others in the society or for society as a whole: precepts concerned with individual freedoms can conflict with the common good. The right to carry a gun in the USA is a well-known example. One may support freedom of expression, but should people be allowed (as a moral and a pragmatic matter) to incite violent rebellion, or to publish recipes for chemical weapons? Freedom of religion is fully accepted in the West, but what if the advocates of one religious system are over-vigorous in voicing their criticisms of another? As I write, a UK politician has questioned the wearing of a scarf that conceals the face by Muslim women. He argued that it made conversation more difficult, as non-verbal cues are not available. The result has been considerable ill feeling amongst Muslims in the country. Again, people should be allowed to demonstrate against taxes of which they disapprove, but in recent years demonstrations against a tax on fuel in the UK came near to paralysing hospitals, schools, and other organizations set up for the public good: the transport minister acknowledged the right to peaceful demonstration, but not to put the livelihood and convenience of other people at risk. Thus while rights in the abstract may be universally recognized within a culture, their definition and application in practice may pose many problems.[18]

Another limit on rights and freedoms is that they are usually seen as applying to individuals with full 'personhood'. Thus denying women the vote is denying them full personhood. Prisoners and the insane are also often seen as deficient in personhood (see below).

So what is the status of 'rights'? It has been argued that no rights exist unless encapsulated in the law. But, if that were the case, a government could create laws that restricted individuals' freedoms in any way that it wished. In practice the situation is the opposite: in most Western countries the law is now judged by moral standards, so that laws that unduly restrict human rights are seen as unjust. This brings with it the hidden danger of too great an emphasis on the inalienability of human rights: if individual rights are what matters, social responsibility can go by the board. Thus justification for a claim of rights must take into account the well-being of the group.

Generalizing broadly, the nature of human rights is seen differently in East and West. When the Universal Declaration of Human Rights was being turned into law, the West argued that civil and political rights had priority, economic and social rights being mere aspirations. The Eastern countries argued that rights to food, health, and education were inalienable, with civil and political rights secondary. As a result, two separate treaties were created in 1996. The difference perhaps reflected the differing need for each category in the East and West.

JUSTICE TO INDIVIDUALS VERSUS GROUPS

We have seen that problems in individual relationships can arise over what is considered to be 'fair'. The same problem arises in relations between individuals and groups. For instance, in relations between men and women, should fairness be equated with equality, or should the differences in attitudes and needs between the sexes be taken into account? Discussions about this issue are often complicated because the matter involves the two distinct yet interrelated levels of individual and group. For instance, if students compete for entry to medical school, and more women are admitted than men, should a less talented man be admitted in preference to a more capable woman, in order to preserve equality between the sexes? In this case, the good of society demands that the more talented individuals be admitted.

CONFLICTS WITH CONVENTIONAL AUTHORITY

All too often individuals are required to act in ways that would be incompatible with their personal morality. In Milgram's[19] famous studies, mentioned already, individuals were required by an experimenter with the trappings of conventional authority to perform actions that they believed were inflicting harm on others. Many complied, acting against their moral principles because of the perceived authority of the experimenter. This can be seen as one moral precept (obey your superior) overriding another (the Golden Rule). These experiments occasioned much surprise at the time, but they need not have done so, for similar scenarios were enacted countless times by the guards of the Nazi death camps and on innumerable other occasions throughout history.

Another source of conflict between individual and societal morality is the time lag in the reciprocal relations between changes in what individuals do and what others believe they should do. We have seen that what individuals do both influences and is influenced by the norms of the society (see pp. 12–13). But an individual may perceive that societal norms are unacceptable before those norms change, or he may be slow to adjust to a changing moral climate. Examples are to be found in the history of opposition to slavery, and the increasing unacceptability of racial prejudices. Again, the suffragettes, campaigning for votes for women, encountered opposition based on rationalizations involving the false presumption of an intelligence difference, implying that women lacked full personhood, yet they did not find universal support even amongst members of their own sex. Comparable problems occurred in the 1920s when some sections of society were emphasizing personal freedom at the expense of the Victorian ideal of selflessness. Variations on this were seen again in the 1960s when personal fulfilment ('Be yourself, man') came into conflict with traditional morality. The changing attitudes to divorce and premarital cohabitation were mentioned in Chapter 1.

INDIVIDUAL VERSUS SOCIETAL
CONVENTION/MORALITY

It is usually considered that moral precepts should take precedence over mere societal convention. However, on occasion some conventions carry moral force. For instance, the speed at which cars should be driven may be limited by a legal convention, so that exceeding the speed limit can be seen as merely contrary to the convention. However, it is considered morally wrong to drive at such a speed that the lives of passengers or others are put at risk. And should loyalty to a small group override the well-being of the wider group? For instance, should you betray a friend who has performed a morally reprehensible crime? Or support a friend who is applying for a job to which you think he is not suited?

Responses to such dilemmas often depend, perhaps regrettably, on whether the transgression is concealed. Thus an individual may feel justified in telling a lie to conceal that he has breached a social convention to avoid ridicule. Conversely and surprisingly, individuals appear to be more likely to help another in distress when alone than when in the company of others who remain passive: perhaps a wish not to stand out from the crowd, or doubts about one's proposed action engendered by the inaction of others, are at the roots of this.

CONFLICTS OVER PRECEPTS OR BELIEFS SEEN TO
HAVE A BASIS IN THE SOCIETAL RELIGION

Religious differences can poison personal relationships in situations that fall far short of causing conflict. The differences may involve beliefs, ritual, or morality. Conflicts over morality are especially likely to be intense with societies where all moral precepts are seen as having a foundation in religion. Differences are likely to be exacerbated when leaders with fundamentalist leanings encourage the establishment of single-faith schools. These can make matters worse.

A less contentious case is the inhibition that agnostics may feel in talking with devout believers. The believers feel offended by the expression of opinions that run contrary to their beliefs, while agnostics feel that their personal integrity is infringed by the necessity to suppress their right to free expression: often the result is a joking relationship, in which neither expresses his/her full opinion.

Regrettably, religious differences often result in group violence, though this is often either because the religious difference is related to a power difference or because it is used by politicians to foment conflict. Disputes seem especially likely between variants on the same central religious doctrine. Antipathy between Catholic and Protestant Christians, originating in differences in employment opportunities and financial status but portrayed as involving a religious difference, has been present for generations in Northern Ireland, and at the time of writing violence between Shia and Sunni Muslims in Iraq is leading to many casualties in the aftermath of the Second Gulf War.

ANTIPATHY TO THE VALUES OF OTHER GROUPS WITHIN THE SOCIETY

As we have seen, during development individuals tend to incorporate the beliefs, precepts, and values of their culture into their self-systems. Just as we see ourselves as Caucasian, or female, or as a bank manager, we also see ourselves as radical, honest, generous, and so on. Thus our basic beliefs and values are part of how we see ourselves, and resistant to modification. But moralities differ between cultures and even between subcultures within a society, and this is frequently a cause of friction. Some may find the values and beliefs of some of their fellow countrymen totally unacceptable, as shown in the past by the horrors of the Inquisition.

Conflict can arise from differences in judgements about categories of individuals within the society, but here it is important to distinguish between judgements of the group and judgements of its individual members. For example, it is easy for middle-class parents to condemn out-of-hand the values of teenagers who 'hang out' at street corners, or worship pop idols. But it is the values of the group that are being disparaged, and the parent should ask whether that is simply because they differ from his own values, or because he can foresee that such values would have bad consequences for child or society if put into practice, or because he perceives that they contravene the Golden Rule. In the first case, at least, an attempt at understanding is desirable before condemnation is justified.

A similar issue arises in retrospective assessments of the bombing campaigns in World War II. One may decide that the bombing was wrong, but that is not to be confused with judgements of the bomber crews themselves. They were carrying out what they considered to be their duty, and showed the most incredible dedication and bravery in doing so (see Chapter 9).

Members of a subgroup tend to limit their prosociality to other members of their own subgroup. The difficulties that arise in relations with members of other subgroups are then particularly problematic. In modern societies, most individuals belong to several groups, and the groups may have different conventions or moral codes. Many find the transition from work to home a little difficult because different conventions apply: one may accept conventions and even moral rules at home that one discards at work (see Chapter 8). Combatants returning home from a war often find the change in convention difficult to cope with—even over the sort of language they were habitually using. Sometimes the conflict is between private and public loyalty, as with the terrible dilemma facing many in times of civil war.

CLASHES BETWEEN WORLD VIEWS

Sometimes conflicts arise from individuals ascribing different meanings to the same moral concept. For instance some, focusing on the group level, see individual freedom as implying unlimited individual competition, whereas others, conscious of the differences between individuals, would see unlimited competition as inevitably leading to the denial of rights to some.

Closely related to such cases are conflicts that arise between people holding different world views. For instance the debates over the acceptability of abortion, contraception, or assisted conception depend on fundamental differences in the ways that people see the world, usually stemming from differences in their religious beliefs.

Where the differences lie between groups they may be exacerbated by false stereotypes of the views of the opposing group, making any form of compromise exceedingly difficult to achieve. A well-studied case concerns the differences between those who accept the scientific view of evolution, and those who accept the biblical story of the Creation—that is, between evolutionists and some religious fundamentalists. Some years ago an analysis of the public discourse of the 'Secular humanists' and the 'Religious Right' found a radical difference between the private intra-mural discourse within each group and their public discourse. The first was characterized by attempts at rationality, intelligibility, and compassion, the latter consisted of reciprocated diatribe. In public discourse each felt its values to be threatened and that it was necessary to oppose the other in every way possible.[20] Discourse between them was virtually impossible.

Such divisions can be even more marked between cultures. For example, inhabitants of Western societies usually see personal freedom, self-actualization, and autonomy as absolute rights—though, as discussed above, there may be limitations in practice, and the smooth running of society demands some constraints on individuals. It has been suggested that these freedoms are related to the Christian deity, which has taken human form and to whom individuals matter: it is assumed that individuals are of high and equal value. This has been contrasted with the emphasis on the group in many Eastern societies. Chinese philosophy stems from the Confucian emphasis on harmony in the universe, with the positions and conditions of people set by the correlations in the universe. Actions are judged according to whether they are seen to be in harmony with the cosmos, and the emphasis is on loyalty to the group. Thus it is wrong to steal because it infringes the integrity of the group, and those who steal become to some extent non-persons, and accordingly are to be dealt with severely.

These characterizations of whole societies are, of course, crude and potentially misleading: they tend to under-emphasize the particular characteristics of other cultures by setting up Western values as a standard for comparison.[21] However, they do indicate that it is hardly surprising that misunderstandings can arise. For instance, for a Westerner it is easy to judge other societies by the degree of freedom allowed to their citizens. To most Westerners (except for many living in the USA), capital punishment has come to seem inhumane, but this is much less the case in the People's Republic of China. Most Westerners regard the position of women in Islamic countries and arranged marriages as unacceptable because they deny women full personhood. Reciprocally, for many living in the societies of the Middle and Far East, Western values have come to be epitomized by fizzy drinks, selfish individualism, cultural imperialism, and lack of spiritual values. It is important here not to be misled by the desire for Western standards of living by those inhabiting countries that are not fully industrialized. They want Western modernization and material goods, but reject Western cultural values.

Cultural differences are often inseparable from religious ones, and religious freedom for a minority group may clash with the perspective of the majority. Such problems have increased in recent years with globalization and the mobility of populations. Often the dispute focuses on symbolic issues, such as the wearing of a headscarf by Muslim women. Some see this as a symbol of Islamic ideology and as implying acceptance of the suppression of Muslim women and the rejection of Western values. Others see it as a consequence of Islamic leaders' attempts to exercise control over individuals independently of the state in which they live. If 'they' come to live in our country, it is argued, 'they' should adopt our values and conventions. On the other hand, some see the prohibition of the headscarf as contrary to the principles of democracy and religious freedom, and as displaying a Western bias because banning the headscarf would be incompatible with acceptance of the wearing of a crucifix by women in the same society.

In other cases disputes arise over the architectural and cultural clash produced by the building of mosques in Western cities, by polygamy in monogamous societies, and differences in religious education. The genital mutilation of women, which has been perceived by at least some of those who undergo the operation as a mark of adult status and full access to their cultural heritage, is condemned by many who accept the circumcision of males as normal. Condemnation of female circumcision may be right for pragmatic reasons, but it should be remembered that there is another side to the case.

Similar differences occur within societies. For example, disagreements over birth control and abortion have led to violence between secular pragmatists and the religious right. The Pope's prohibitions may be based on a religious tradition, but

to outsiders (and increasing numbers of Catholics) the consequences in terms of individual freedoms, the rights of women, the
problems of world population, and the spread of HIV/AIDS are
horrific.

From the perspective on morality presented here, the most
serious of these conflicts are those that contravene the Golden
Rule and affect the personhood of individuals. In modern societies personhood and its constituent rights are denied to individuals deemed to be insane and to criminals. This is justified in
terms of benefit to society. But formerly in our own society, and
currently in some others, slaves were considered non-persons. In
India it has been a matter of birth, certain individuals being considered as outside full society. In fundamentalist Islamic regimes,
women are denied their rights as citizens and are effectively non-
persons. Such practices are inevitably seen by those outside the
culture as infringing the pan-cultural Golden Rule.

I have mentioned already the even more difficult question of
judgements of those who live in a system that we see as wrong,
but in our view 'should have known better'. This is the central
issue in many war crimes. Consider an individual, brought up
in what we would consider a moral manner, indoctrinated and
employed as a concentration camp guard ordered to carry out a
genocidal policy. Leaving aside the question of whether he was in
danger of punishment for not obeying orders, should we exoner-
ate him as acting within the system in which he now believed, or
blame him because he should have known better?

In general, is the proper course to attempt to eliminate cultural
differences? In the first place, that would be an unattainable goal.
As indicated above, cultural values are deeply embedded in the
self-systems of individuals, and could be changed only over generations. Attempts to convince members of another culture to
conform to one's own values by military, political, or economic
means are unlikely to be successful. The present US-led efforts

in this direction in Iraq have exacerbated the divisions in the world, encouraged the spread of Islamic fundamentalism, and led to greater violence between religious groups than there was before. In any case, we may ask if a uniform Coca-Cola world would be desirable: we should treasure the glory of local diversity. And the missionary practice of conversion amounts to an assumption of moral superiority and treating outsiders as ignorant savages. Rather we must value what is best in our own cultural traditions, cutting out its undesirable excesses, and be open to criticisms from elsewhere. We must refer constantly to the universals of human nature, enhancing the propensity for prosociality and respect for personhood and the Golden Rule and the precepts that follow from it. We must try to improve our own culture and to understand the cultures of other societies, and perhaps their members will see that the way of life that we have developed brings dividends in human fulfilment and well-being. Or, one must concede, one may find the comparison painful.

CONCLUSION

In the first two chapters I considered how morality develops in the individual, what causes an individual to behave morally in terms of the maintenance of congruency in the self-system, the biological evolution of morality, and, by implication, the function of some of our moral precepts. In Chapter 3 I supplemented this by speculating about the probable cultural evolution of morality and its relation to law. I suggest that this four-pronged ethological approach[22] is necessary if we are to comprehend the nature of morality, and that knowledge of how and why we have the moral principles and precepts that we do have helps to tell us why we should (or should not) follow them.

This chapter has been concerned with some everyday issues relevant to the thesis of this book. Exchange theories throw light on many aspects of personal relationships, including the role of prosocial reciprocity. The Golden Rule is basic to at least the great majority of all moral codes, and many human values and virtues spring from that. But the relationship between one individual and another cannot usefully be considered in isolation from their relationships with others in their group. I have also emphasized that moral conflict is nearly always potentially present, even in personal relationships. Most of the time we are not conscious of any conflict because the precepts incorporated into our self-systems steer us in the right direction.

The following chapters have two aims. First, I aim to show how a sample of contexts of modern life almost inevitably impose moral conflicts on those involved. Second, I suggest some ways in which people come to terms with breaking moral rules. Let us look ahead and see where this takes us.

For a group to function effectively, the individuals must cooperate with each other. But if resources are scarce, they must also compete. An individual who adhered rigidly to the moral precepts might well lose out in competition with individuals who did not. In addition, if precepts were rigid and rigidly followed, many conflicts would be impossible to resolve and the resulting tension would be unbearable. But what matters for the reduction of personal tension in a relationship is how the actor perceives the situation, not whether it is 'actually' fair or good or proper. If you have imagined yourself in the conflict situations discussed in this chapter, have you been surprised by how often you are sure that your criterion of what is fair is the right one, without considering the other's perception? How often have you used dubious premises to justify what you do, or seen your own needs to justify your actions? Have you justified taking more than your share because you think you need more? How often have you seen duty to friend, group, church, or country override your moral scruples? Have you just managed to find a good reason why the course of action that you contemplate is just and proper? Have you ever managed to perceive that deceiving a friend is in his interests, *really*?

I believe that there are few who would not answer 'yes' to most of these questions, and of course I include myself here. But I am not simply bewailing the infirmity of the human condition. We need moral rules to guide our behaviour, to maintain the balance between self-assertiveness and prosociality, and we need to perceive those rules as absolute. If we did not see them as absolute, social life would be endangered. Aligning our actions with the rules incorporated in our self-systems to preserve congruency is a part of human nature (in the strict sense): it makes group-living possible. But every positive answer to the questions I have posed shows that we can often find ways round the rules. We need a little flexibility because the perfectly moral person would soon be exploited by others. And if we perceive ourselves to behave in ways that conflict with the precepts incorporated in our self-systems, we can make adjustments to restore congruency by amending how we perceive our actions, or how we perceive the rules, or how we evaluate the opinions of third parties (see pp. 25–7). Over minor issues, perhaps that is not a bad thing. Life is complicated, consciously or unconsciously we are frequently making moral decisions, some conflicts are almost insoluble, and guilt or remorse cannot be sustained for ever.

Does it sound shocking to say that we can sometimes bend the rules and not notice that we are doing so? How often have your own needs taken precedence over duty to friend, group, or country? That is not to say that we should not perceive the rules as absolute. Nor to say that we should not try to abide by them. It is only pointing out that we sometimes need to get round them. It sounds anomalous, but maybe that is the only way in which morality can be maintained in a group: if those with honest intentions could not get round the rules sometimes, they would soon be exploited by the free-riders.

But, and it is a big BUT, if these manipulations of conscience got out of hand, society would disintegrate. We try to take care of this with a legal system which, to be effective, must itself bend the rules. But the real danger comes when they become more or less accepted and institutionalized in particular areas of social life, for morality may then be forced to the edge. The following chapters concern contexts in which this is happening to various degrees.

5

Ethics and the Physical Sciences

INTRODUCTION

This chapter and the next are concerned with ethical problems that arise in the physical and medical sciences.[1] The physical sciences are discussed first because ethical problems there have become acute only relatively recently, and there is as yet no clear plan on how they should be met. Ethical problems in the practice of medicine and in medical research are well recognized, and there are methods for dealing with them—though the rapid advances in medical and biological sciences are bringing new and even more acute problems. Later chapters deal with contexts in which the problems have long been with us, but those involved, or the community, have devised methods to reduce the tension that they would otherwise generate.

Pure science strives for objectivity through such processes as peer review, the replication of experiments, and the widespread dissemination of its results. Nevertheless, in recent years ethical issues have come to pervade research in the physical sciences both in the details of its practice and in the responsibility of the scientist for its consequences. Differences of opinion arise as to whether scientists should be concerned with issues involving the social impact of their work, or whether the pursuit of knowledge should be their sole guiding principle. Should they accept responsibility for the human and environmental consequences of scientific research? Science has brought untold benefits to humankind, but it has also given us weapons of mass destruction—nuclear, chemical, and biological. Applied research presents its own problems—problems that can seem both intractable and urgent because of their immediate relevance to human life.[2]

A HISTORICAL PERSPECTIVE

Questions about the ethical responsibility of scientists did not arise in the early days of science, because scientific research had few if any consequences for human welfare or the environment. In those days science had no role in the day-to-day life of people, or, with a few exceptions such as Archimedes and Leonardo da Vinci, in the security of states. Even in the seventeenth century, when the Royal Society (the national academy of sciences of the UK)[3] was founded, science was largely the pursuit of a few gentlemen of leisure who collected plants or fossils, gazed at the sky and noted unusual events, or performed simple experiments. There were no journals and no internet, and so they communicated their observations to other gentlemen with similar hobbies at gatherings of a social character, a sort of salon entertainment. The impulse for those pursuits was primarily sheer curiosity, with perhaps an element of competitive status-seeking over their fellow scientists, not unlike most scientists today. The desire for useful applications was present, but it was not the main aim.

In course of time, science began to be taken up as a full-time profession; learned societies and academies of science were established, with highly exclusive memberships, and this increased even further the detachment of scientists from society. One of the founders of the Royal Society,[4] the famous physicist Robert

Hooke, stipulated that the Society 'should not meddle in Divinity, Metaphysicks, Moralls, Politicks, Grammar, Rethorick or Logick'. This detachment of scientists from general human affairs led them to build an ivory tower in which they sheltered, pretending that their work had nothing to do with human welfare. The aim of scientific research, they asserted, was to understand the laws of nature: these are immutable and unaffected by human reactions and emotions. Scientists saw themselves as an in-group, with responsibilities to each other but not to the outside world (pp. 46–8).

Arising from this exclusivity, scientists evolved certain precepts about science to justify the separation from reality. These precepts included: 'science for its own sake'; 'scientific inquiry can know no limits'; 'science is rational and objective'; 'science is neutral'; 'science has nothing to do with politics'; 'scientists are just technical workers'; 'science cannot be blamed for its misapplication'. Such attitudes are now unreal: scientists are part of society and their work affects and is affected by the social context in which they live.

The ivory tower mentality was perhaps still tenable in the nineteenth and even the early twentieth century, when a scientific finding and its practical application were well separated in time and space. The time interval between an academic discovery and its technical application could be of the order of decades, and it would be implemented by different groups of scientists and engineers. Pure research was carried out in academic institutions, mainly universities, and the scientists employed in these institutions usually had tenure: they were not expected to be concerned about making money from their work. The taking out of patents occurred very seldom and was generally frowned upon. This enabled academic scientists working in universities to absolve themselves from responsibility for the effects their findings might have on other groups in society.

The scientists and technicians who worked on the applications of science were mainly employed by industrial companies whose chief interest was financial profit. Ethical questions about the consequences of applied research were seldom raised by the employers, and the employees were discouraged from concerning themselves with such matters.

All this has changed. The tremendous advances in pure science, particularly in physics during the first part of the twentieth century, and in biology during the second half, have completely changed the relationship between science and society. Science has become a dominant element in our lives. It has brought great improvements in the quality of life, but also grave perils: global warming, pollution of the environment, squandering of vital resources, global transport leading to the spread of transmittable diseases, and above all, a threat to the very existence of the human species on this planet through the development of nuclear weapons.

An important outcome of the change of emphasis in scientific research is the narrowing of the gap between pure and applied science. In some areas the distinction has become very difficult to discern. What is pure research today may find an application tomorrow and become incorporated into the daily life of the citizen next week. In many areas of research scientists can no longer claim that their work has nothing to do with the welfare of the individual or with politics. The ethical principles guiding the scientist must include recognition of his or her social responsibility.

Unfortunately some scientists do claim that they have no responsibility for the application of their research. Clinging to the ivory tower mentality, they still advocate a laissez-faire policy for science. Their logic rests mainly on the outdated distinction between pure and applied science. It is the application of science that can be harmful, they claim. So far as pure science is concerned, the only obligation on the scientist is to make the results of research known to the public. What the public does with them is its business, not that of the scientist. While the distinction between pure and applied science is still clear in some areas of the physical sciences, in others it is largely non-existent, and in the latter such an amoral attitude is unacceptable. Indeed, it is an *immoral* attitude, because it eschews personal responsibility for the likely consequences of one's actions.

We live in a world community with increasing interdependence between individuals and between nations; a trend due largely to technical advances arising from scientific research. An interdependent community offers great benefits to its members, but by the same token it imposes responsibility on them. Every citizen has to be accountable for his or her deeds: we all, and this includes scientists, have responsibilities to our peers. Indeed, this responsibility weighs particularly heavily on scientists precisely because of the dominant role played by science in modern society. Scientists understand technical problems and probabilistic predictions better than the average politician or citizen, and knowledge brings responsibility. While their main purpose is to push forward the frontiers of knowledge, this pursuit should contain an element of prosocial utility, that is, benefit to the human community. This means giving some precedence to projects likely to advance the welfare of humankind and the environment, and a total ban on those likely to do harm, while bearing in mind the enormous difficulty of predicting the long-term consequences of any investigation. The ethical principles guiding the scientist must recognize his or her social responsibility. A statement made nearly four hundred years ago by Francis Bacon is fully applicable to the present time.

I would address one general admonition to all: that they consider what are the true ends of knowledge, and that they seek it not either for pleasure of the mind, or for contention . . . but for the benefit and use of life ... that there may spring helps to man, and a line and race of inventions that may in some degree subdue and overcome the necessities and miseries of humanity.

If Bacon were speaking today, he would probably add: '... and to avert the dangers to humanity created by science and technology'.

SOME GENERAL ETHICAL PROBLEMS IN SCIENCE

Scientists are human beings, and subject to human failings. In particular, it is possible for a scientist to cheat, especially if he or she can be fairly confident that no one else will know. Thus it is sometimes tempting to suppress data that fail to replicate earlier findings, to cook results, or to plagiarize another's work. Fortunately these things are not common. Most journals ensure that the papers they publish have been submitted to review by two or more referees, who should be in a position to detect malpractice. But ultimately it is a matter for the individual's conscience and, though there have been exceptions, most scientists feel strongly about the integrity of the scientific enterprise.

Many of the problems that do arise come from the hierarchical structure of most research institutions. Past progress in research often leads to increased specialization and teamwork. The team leader not only has responsibilities to the funding agency and to the public to push the research forwards, but also to ensure the fair treatment of team members. It is highly desirable, for instance, that young scientists should be given the opportunity to show originality, and that they should receive credit for their work. These issues affect particularly research and postdoctoral students. Should the research supervisor put his name on papers that result from a student's thesis? Authorship means much to the young scientist, and may be important for his or her future career. How much independence should be given to postdocs? The answers to such questions depend on the nature of the research, making generalizations difficult, but authorship should depend on the contribution made to the research. In addition, the good supervisor must recognize that authorship means more to the student than it does to him. The duty to train students is not satisfied by merely giving them a topic and telling them to get on with it.

Partly as a consequence of the virtual disappearance of the boundary between pure and applied science in some areas, research has become motivated by financial gain. Although this has been somewhat less of a temptation for the scientist working in a university or research institute, financial considerations are causing more and more such institutes to require their employees to take out patents and share the profits with their parent institute. Frequently, this works against one of the main postulates of scientific research, namely that the results of research should be available to everybody, to be used for the public good. An important example is the Bermuda agreement by scientists working on the human genome not to patent their results but to pool them centrally as the research proceded: this accelerated progress in this vastly complex enterprise. But scientists, or their employers, are often caught between scientific integrity and the ethics of the marketplace (see Chapter 8). For instance, the financial promoters of research projects, particularly in the pharmaceutical industry, may impede the publication of findings, either prohibiting publication altogether or adding considerable delay, in order to further their own interests in competition with other firms.

The whole practice of patenting scientific findings not only goes against a basic tenet of science, but also affects the pursuit of science by exacting payment for the use of essential materials and for the technologies covered by patent rights. The granting of patents for certain results of scientific research, particularly research on basic materials such as genes, is surely not acceptable. At present natural phenomena or laws cannot be patented, but a process that takes advantage of them can be. A more difficult question concerns whether a biological relationship can be patented. A US Supreme Court decision allowed a microbe capable of digesting petrol to be patented. Many think the gates are now opened too wide.[5]

Nearly every university now receives some research support from commercial undertakings. Reciprocally, the commercial organizations profit from the expertise and facilities available at universities. Both profit in terms of employment for scientists and in terms of the contributions to knowledge that result. On the other hand, as will be apparent from Chapter 8, corporate interests are often served by secrecy, by the establishment of individual property rights, and by a narrowing of the focus of research. The counterclaim is that financial gain to a commercial enterprise from scientific research is necessary to finance further research, but it is not acceptable that either progress in science or its public benefits should be limited by commercial considerations. Corporate interests easily tend to detract from the proper nature of scientific research.

Secrecy in scientific research for the financial profit of a commercial company is only one aspect of a multifaceted problem. Another is secrecy imposed by scientists themselves to safeguard against other scientists stealing their ideas, techniques, or results. Both the plagiarism and the secrecy it induces are ethically indefensible. My conviction that, in the early days of my scientific career, the excitement of the joint enterprise and the multiplicity of the problems before us permitted much greater openness, is not just the 'Good Old Days' syndrome. Now increasing specialization and the way science is organized has brought a tendency towards secrecy until the work is accepted for publication. The increasing emphasis on primacy in publication tempts scientists to conceal from each other the techniques that they are using, or the progress they have made, so that others would not profit from their work. First past the post gets all the credit, no matter how meritorious the work of others who later achieve the same goal.[6]

The prestigious and financially valuable prizes that are available in some fields, and the exclusiveness of scientific academies, have exacerbated the situation. Of course, while the possibility of financial reward is relatively new, the phenomenon of competition is not. Darwin was deeply disturbed when he learned that, before he had published his theory, Wallace had come to a similar conclusion. So here is a dilemma. Competition acts as a spur to progress, the acclamation given to individuals of exceptional ability acts as an example to others, and the academies must maintain the highest standards so that they can promote scientific research. The downside is that individuals are less often content with adding their brick to the edifice of science, less often concerned with the prosocial pursuit of truth for the common good, and motivated more by selfish assertiveness. In addition, scientific awards tend to be given to individuals who are on the near-flat part of the bell curve of academic merit, where there are a number of top-class candidates, so that the decisions of selection committees, though reached in good faith, have a degree of arbitrariness. Here are problems not easy to resolve but which scientists have to tackle.

There are strong arguments against any restrictions on research in pure science. The freedom of scientists to pursue their own interests is in itself important, and conducive to scientific progress. Any advance in understanding the nature of the universe or of ourselves helps us to gain a more valid perspective on the nature of the world and of our lives and relationships in it. The contributions of pure science to human welfare have been immense, and we shall never know when we know all there is to know in a particular field, or even everything that is worth knowing. But pure science has also provided knowledge that has been to the detriment of humankind, and discriminating between ethically acceptable and unacceptable research projects is often extremely difficult if not impossible. On the one hand, who would have predicted what benefits the discovery of X-rays would make possible? On the other, no one could know that efforts to understand the nature of matter would lead to the nuclear bomb. It is the application that matters but, as we have seen, the border between pure and applied science has become hazy in some areas, and the distinction may no longer be easy to make. Indeed, it is usually not a matter of deciding whether a given research project is or is not acceptable, but of weighing potential benefits against potential harm when benefits and harm are measured in different currencies. Though prediction is inevitably hazardous, no one is better qualified to judge which way the balance will tip than the scientists themselves. Unfortunately, while most scientists are now conscientious, the ivory tower mentality persists among some. Indeed it sometimes seems as though the ivory tower is an excuse constructed by such scientists to give them free licence in their research.

WEAPONS RESEARCH

The issues of secrecy and financial gain arise with special urgency in research on weapons. Existing in a competitive ethos, armament firms are constantly striving to produce weapons that are more deadly or more efficient than those currently available. To that end, scientists and technologists devote their time to inventing or perfecting ways to kill. Of course, this is an old problem, going back to the invention of the crossbow and beyond.[7]

Nowadays an especially serious problem is posed by the many thousands of scientists still employed in research on the development of new, or improvement of old, weapons of mass destruction in national research laboratories, such as Los Alamos or Livermore in the USA, Chelyabinsk or Arzamas in Russia, and Aldermaston in the UK. What is going on in these laboratories is not only a terrible waste of public funds and scientific endeavour but a perversion of the nature of science. Rather than taking part, scientists should oppose it.

The Nobel Laureate Hans Bethe, one of the most senior living physicists, and one-time leader of the Manhattan Project which produced the first nuclear weapons, said:

Today we are rightly in an era of disarmament and dismantlement of nuclear weapons. But in some countries nuclear weapons development still continues. Whether and when the various Nations of the World can agree to stop this is uncertain. But individual scientists can still influence this process by withholding their skills.

Accordingly, I call on all scientists in all countries to cease and desist from work creating, developing, improving and manufacturing further nuclear weapons—and, for that matter, other weapons of potential mass destruction such as chemical and biological weapons.

This call should be endorsed by the scientific community. All scientists should demand the elimination of nuclear weapons as both illegal and immoral and, in the first instance, request that the nuclear powers honour their obligations under the Non-Proliferation Treaty (see Chapter 9). Unfortunately, at the moment movement in that direction is blocked by the intransigence of certain states, especially those that already possess nuclear weapons, and the hegemony of the policies of the US administration.

The clearest danger results from the fact that some techniques developed in the scientific laboratory or used in industrial processes are readily converted for use in the manufacture of weapons of mass destruction—the so-called 'dual-use' problem. This gives rise to a major difficulty in verifying international treaties forbidding the manufacture of weapons of mass destruction. The legitimate preparation of slightly enriched uranium for use in power stations is a first step in the preparation of the highly enriched uranium necessary for nuclear weapons. The International Atomic Energy Agency has to differentiate between the two. A different problem arises with chemical and biological weapons. Not only are the facilities required for their preparation more difficult to detect, but research aimed merely at the protection of soldiers and civilians from chemical and biological agents may require preparation of the agents themselves. This poses major problems for verifying international treaties forbidding their use. Nevertheless, there is a curious imbalance in that research on chemical and biological weapons is forbidden, but research on nuclear weapons is not.

But what about the individual scientists engaged in weapons research? Most of them may not see their mission as perfecting ways to kill. A few may be motivated by considerations of national security, but the vast majority are lured into this work by the siren call of rapid advancement and unlimited opportunity to pursue research questions. Theodore Taylor, one of the chief designers of the atom bomb in Los Alamos, recalled:

The most stimulating factor of all was simply the intense exhilaration that every scientist and engineer experiences when he or she has the freedom to explore completely new technical concepts and then bring them into reality.

Whatever their motivation may be, whatever the reason they give themselves for engaging in weapons research, the fact is that they are contributing to killing and all the suffering that war entails. They have downgraded the Golden Rule, though probably many do not see themselves as having done so, they just have not thought through the ethical implications of what they are doing. That points to the need for the proper education of scientists.

ETHICAL CODES FOR SCIENTISTS

Weapons of mass destruction, or indeed weapons of any sort, are not the only problem. For instance, another example of current interest is nanotechnology, which involves the manipulation of materials at the atomic, molecular, and macromolecular scales. This is leading to the development of new materials, and there are promises of applications in electronics, medicine, and other fields. What is not yet known are the dangers inherent in very small particles, the problem being exacerbated by the fact that most of the work in nanoscience is not in the public domain. There is a clear need for openness and for regulation in this field.[8]

The application even of more traditional science may bring destruction of the environment, global warming, shortages of basic natural materials, and so on. The misuse of science is in part due to its marriage to a capitalist economy, and we do not yet know how misuse can be prevented. One route that is being tried is to express the desiderata in an ethical code of conduct for scientists, perhaps formulated in some modernized form of the Hippocratic oath formerly taken by medical practitioners (see Chapter 6). Whereas the necessity for ethical principles to be exercised in medicine has been recognized for hundreds if not thousands of years, such recognition has come more recently in the physical sciences. Nowadays, however, the doctor's relationship to the patient is paralleled by the scientist's relationship to the whole of humanity, and indeed to the whole biological and physical world. It is now highly desirable to frame codes of conduct, suitable for specific areas of research, to ensure that scientists do not engage in research that they believe may have consequences harmful for humankind.[9] It may be that the time has come for scientists in some areas to be registered and to be subject to ethical precepts limiting the sort of research they undertake, as is already the case with medical doctors and psychologists concerned with human subjects. That will take time to put in place in other scientific disciplines: in the meantime some have suggested that there should be some kind of oath, or pledge, to be taken by scientists, perhaps when receiving a degree in science.

Three kinds of codes of conduct can be distinguished: Aspirational codes, designed to stimulate reflection on moral issues; Advisory codes, often specific to a particular area, giving guidance to scientists; and Disciplinary codes, ranging from international law to moral clauses embedded in research contracts.[10]

Various formulations of oaths to suit specific conditions have been suggested and introduced by some professions. Some are long and detailed, some short and general. An example of the latter is the oath adopted by the European Physical Society:

In all my scientific work I will be honest and I will not do anything which in my view is to the obvious detriment of the human race.

If, later, I find that my work is being used—in my view—to the detriment of the human race, I will endeavour to stop these developments.

In this formulation, the scientist undertakes not only to refuse to work on projects harmful to society, but to take action to stop such research going on. A somewhat different approach is taken in the oath for Scientists, Engineers, and Technologists, suggested by the Institute for Social Inventions:

I vow to practise my profession with conscience and dignity;

I will strive to apply my skills only with the utmost respect for the well-being of humanity, the earth and all its species;

I will not permit considerations of nationality, politics, prejudice or material advancement to intervene between my work and this duty to present and future generations;

I make this Oath solemnly, freely and upon my honour.

A formulation suitable for young scientists to be taken at graduation has been adopted by the Student Pugwash Group[11] in the United States. This Pledge, already taken by thousands of young scientists in several countries, reads:

I promise to work for a better world, where science and technology are used in socially responsible ways. I will not use my education for any purpose intended to harm human beings or the environment. Throughout my career, I will consider the ethical implications of my work before I take action. While the demands placed upon me may be great, I sign this declaration because I recognize that individual responsibility is the first step on the path to peace.

However, there is a problem with such formulations.[12] They could place some young scientists in an impossible moral dilemma, especially if employment opportunities should be rare. Scientists with young children might face a very difficult decision if they had doubts about the propriety of their work. More importantly, problems arise from the unpredictability of the consequences of particular lines of research, and from the fact that a commercial firm offering employment may be engaged in many lines of research, only some of which are harmful. Furthermore, the making of promises in a situation in which a proportion are likely to be broken downgrades the nature of a promise. It might be preferable for the statement to be framed as an undertaking or statement of strong intent rather than as an oath. It would then have an important symbolic value, and would generate awareness and stimulate thinking on the wider issues among young scientists.

If scientists are to be conscious of and meet ethical demands, some form of ethical education is essential.[13] University curricula should, but at present rarely do, include courses of lectures on the ethical aspects of science. But it is also important not only that new entrants into a scientific career should become aware of their social responsibilities, but also that senior scientists should acknowledge their own awareness of such responsibilities. For this purpose, national academies of sciences (or corresponding bodies in countries where there are no academies), should explicitly include ethical issues in their terms of reference. The charters of some academies already contain clauses that allow them to be concerned with the social impact of scientific research. These clauses should be made mandatory; and there should be explicit statements that ethical issues must be an integral part of the motivation of scientists.

To add teeth to these measures, procedures similar to those described in the following chapter for medical and psychological research should be extended to the physical sciences. Committees, preferably composed not only of scientists, should examine the ethical consequences and the potentially harmful long-term effects of proposed research projects. The ethical committees should work under the auspices of the national academies of sciences in the country, but it would be essential also for the criteria used in the assessment of projects to be agreed internationally, so that the same standards are applied everywhere. The International Council for Science (ICSU) seems to be the appropriate organization to coordinate the task. In some countries ethical vetting is already carried out by formal or informal bodies, but there is the need for general acceptance and for an implementation mechanism, and this is where the ICSU should come in.

Of course, it is important not to get things out of proportion. The terrible power of nuclear weapons tends to dominate discussion of these issues, though most research in the physical sciences is likely to be beneficial or neutral to humankind. Although nuclear, chemical, and biological weapons are not the only harmful consequences of scientific research, ethical committees should be able to deal with most applications quite quickly and only a few will require detailed consideration.

WHISTLE-BLOWING

Scientists should go further than ensuring that they are not involved in activities to the detriment of humankind. Their knowledge often enables them to foresee what activities are likely to be harmful. It is then their duty to make their views known in the appropriate circles. For example, not only should scientists not work on nuclear weapons, but it is their duty to bring the nature of these weapons to public knowledge.

If, in the course of work in a governmental, university, or private situation, scientists become aware of an undisclosed application that is being developed which could result in harm to society, it should be their duty to disclose it to relevant authorities, or, if need be, to the general public, so that steps can be taken to prevent the damage. Indeed, whistle-blowing should become part of the scientist's ethos.

As we shall see also in the business world, this practice is dangerous for the whistle-blower, making them subject to reprisals, dismissal from their jobs, or even more severe punishment. The most extreme case is that of Mordechai Vanunu, the Israeli technologist. When he found out that the Dimona plant where he worked was actually engaged in a clandestine project to produce plutonium for nuclear weapons, he resigned from his post, and brought to the notice of the world the true situation. For this he was sentenced to eighteen years' imprisonment, most of it in solitary confinement. Even when he had served this unjust sentence, the Israeli authorities placed severe restrictions on his movements. There is an urgent need for legislation to provide immunity to whistle-blowers.

DISSEMINATION OF SCIENCE

Scientists have a duty to the community in which they live, and which in many cases supports their work. The dissemination of research results is desirable not only for the further progress of science, but because the public has a right to share them. Support for research in universities comes largely from the public purse, and the taxpayer has a right to know how the money is used.

A recent report of the Royal Society emphasizes the importance of furthering public understanding of the scientific issues of the day, especially those that affect public well-being and safety; ensuring that scientists are accountable for what they do; and helping the public to make informed decisions over issues that affect their lives.[14] Since 'informed decisions' often depend on assessing relative probabilities, there is also a need for public understanding of the nature of such decisions. Furthermore, sometimes it is incumbent on scientists to intervene in the public domain. When the SARS epidemic occurred, false rumours that it had come from outer space and that it had been accidentally released from a weapons laboratory in China were widespread, and there was insufficient scientific comment. An important duty of scientific academies is to issue statements on matters of public concern, such as the long-term storage of nuclear wastes and renewable sources of energy.

CONCLUSION

At the beginning of this chapter we saw that science started as a hobby rather than as a profession. That scientists now earn their livings by practising science inevitably raises the possibility of abuse and the need for ethical guidance. Competitiveness has positive consequences, but it also opens the way for the selfish assertiveness of individual scientists to deviate from ethically desirable behaviour. Although rare, plagiarism and secrecy are a continuing problem. An important counterforce to malpractice has been the perception by scientists that they are a group with common values, perceiving science as an institution, seeking after truth, with a duty to a common enterprise. Scientists are proud of their calling and are rarely willing to sacrifice their integrity. Pride in their calling leads scientists to maintain their own integrity and to scorn free-riders.[15]

The professionalism of science does, however, increase their perceived independence from society. The major problem lies with predicting how research will be used. Scientific research involves pushing into the unknown, and whether the results of research will be beneficial, neutral, or harmful with respect to human welfare is often difficult or impossible to predict. So far, there has been too little attention to public opinion on the ethical or financial aspects of research in the physical sciences. This is partly because the research and its possible ethical implications cannot easily be understood without a great deal of background knowledge which few non-scientists possess. Even with such knowledge, and even more without it, the relative value of a new accelerator measured in terms of what the money would do in terms of, say, new hospitals can not easily be assessed.

Whether or not to engage in a given research project is a decision that tends to rest with the scientists themselves, and they are liable to be biased by their own speciality. The specialist training of scientists leads them to believe that they know best. Most scientific research is extremely unlikely to have undesirable consequences. But one can never be sure, and in some cases the results of research can be predicted reasonably well: research into weaponry will lead to weapons that are more efficient instruments for killing. Those who engage in such work may tell themselves that how the weapons are used is no concern of theirs, using the myth of the ivory tower to legitimize their work. They see fellow scientists as an in-group: the rest as outsiders. Or they may justify it as necessary for the defence of their country, seeing duty to country as overriding everyday morality. Most probably, many of them simply do not envisage the human consequences of their work. Some of the blame for this situation lies with the inadequacy of the education they have received.

Unfortunately, especially when scientists are employed by commercial or governmental organizations, their livelihood or advancement sometimes prompts them to forget about the consequences of the research they are conducting. Even if a scientist sees that the research may lead to undesirable consequences, it is difficult for the individual to fight against this, except by whistle-blowing and seeking for employment elsewhere. Furthermore, the dilemmas that arise often appear in black and white terms, when actually it is a case of weighing desirable and undesirable consequences against each other when neither is accurately predictable. The task is perhaps more difficult than in medicine, in part because the problems have been recognized more recently and in part because the relation of the scientist to the community or to future generations is more intangible than that between doctor and patient. On the bright side, we have seen that scientists are becoming increasingly aware of the possible misuse of science, and scientific academies are devising methods whereby research with deleterious consequences can be avoided.

In conclusion, the basic human value is life itself; the most important of human rights is the right to live. It is the duty of scientists to see to it that, through their work, life will not be put into peril, but will be made safe and its quality enhanced.

6

Ethics and Medicine

The development of poisonous gases, nuclear weapons, and other twentieth-century scientific advances has brought the ethical problems associated with research in the physical sciences into prominence only relatively recently. Ethical issues associated with the *practice* of medicine have been a matter of public concern for millennia and biomedical *research*, while opening up possibilities for curing previously incurable diseases, is also raising new and important ethical problems. Opinions differ as to whether some of its applications are desirable. Public opinion, which may include strongly held but contradictory views, views that may not be fully informed, must be taken into account. Nearly always it is a probabilistic matter of weighing possible benefits against possible harm. The medical research worker is better informed than the public and may feel that he knows what will be for the general good in the long run, but the public may not accept his view, seeing him as caught in his own perceptions of the importance of his work. We have seen that our moral precepts are largely maintained or modified according to the degree of public acceptance, and that is equally the case with medical research. Laws relating to medical research do and should involve constraints imposed by public opinion. Such laws are necessary in case the authority of science should be extended to domains that are properly the concerns of all citizens,[1] but carry the danger that essential medical knowledge will be ignored in favour of democratic decision-making.

With the medical practioner, a similar problem arises. Given his long training, he certainly knows better than the patient. But does that entitle him to impose his view on that of the patient?

MEDICAL RESEARCH

Progress comes with new treatments or new drugs. In most cases their use on patients must by law be preceded by both trials with animals and clinical trials, but trials in themselves involve ethical problems. Clinical trials mostly concern the rights of the individual subject set against the (long-term) good of the community as seen by the research worker.

A clinical trial necessarily involves a control group, which receives a placebo, and one or more experimental groups which receive treatments. At least four ethical issues arise from conflict between the rights of individual participants and the probable consequences of the research for treatment. First, it would be unethical to give any of the experimental groups or the controls a treatment that was deemed to be less effective than the treatment currently in general use. The effectiveness of the experimental treatment, however, may not be known with certainty until the trial is completed, though the trial should not be undertaken if there were no expectation that it would give positive results. Second, the subjects chosen for the trial must be representative of the population with whom the treatment is expected to be used. This may rule out the use of volunteers, or the use of a financial incentive for participation that would be more attractive to some than to others. Third, subjects must be allotted to the experimental and control groups in a manner that will not bias the results: this usually means random allocation to groups. Finally, subjects must consent to participating in the trial, and be fully informed as to what it involves and any risks it is known to entail. For instance, participation may mean that a subject is placed in the control group, with no expectation that the treatment will be successful. (Sometimes this can be circumvented by giving both groups both treatments in a cross-over design.) Although informed consent is ethically essential, if it leads to significant numbers of potential participants refusing to take part, the trial may be invalidated as unrepresentative of the target population.

When trials have been concluded, further problems arise. If the preparation is likely to be of enormous help to millions but has deleterious effects on a few, should it be released for general use? How far should limited resources, that could be spent on other aspects of medical care, be spent on expensive treatment for an uncommon complaint? This dilemma may be even more acute in poorer countries.

Other problems arise over the use of medical records or stored samples for research. Present regulations in the UK require informed consent for every purpose for which the records or samples are to be used. As recent distressing cases in the UK have shown, the storage of body parts 'in case' they might be useful later must be ruled out, in part because of the possible distress caused to relatives. While ethically proper, this can be a real brake on research, as new possibilities for the use of stored samples may arise over time.[2]

The moral problems that arise in medical research are diverse, and it is impossible to refer to them all. A few specific examples follow.

The first concerns problems that arose early on in research on the treatment of terminally ill patients with HIV/AIDS.[3] Situations arose where there was a conflict between the use of experimental treatments and the welfare of the patients. Even with known procedures a balance had to be found between treatment and palliation. New treatments had the possibility of bringing great benefits to future patients, but what if the treatment's effectiveness for particular patients were dubious? How far should an aggressive treatment, with short-term deleterious effects, be continued in the hope of long-term benefit? Here the ethical conflict involves the welfare of the individual patient in comparison with that of both future and present generations, for the treatment of HIV extends the life of patients and thus also their potential for infecting others.

A related issue concerns the use of antiretroviral therapy for HIV/AIDS. Once started, therapy must be completed so that HIV does not develop resistance to a variety of drugs. Should the treatment be given, when there is a risk that many sufferers do not complete the course?

The next two examples involve not only doctor and patient, but a clash of world views that seem almost impossible to reconcile. Consider first the case of *in vitro* fertilization. The chairperson (Warnock) of a UK committee, set up in the nineties to consider what legislation is appropriate, argued that it involved public opinion, practicality, and the law.[4] At the time, *in vitro* fertilization was necessary for research on human embryos and the causes of infertility. The committee felt that infertility is a cause of great distress to many. However, many felt that the use of a human embryo for research, with its subsequent destruction, involved the deliberate taking of a human life, and was therefore morally wrong. The issue was whether, or when, an embryo becomes a human individual. There were wide differences of opinion—conception, a certain cell stage, brain activity, viability with minimum support, or birth. The committee eventually chose fourteen days from fertilization, at which point the embryo consists of an assemblage of undifferentiated cells, each of which could form any part of an embryo. After fourteen days the 'primitive streak', which will eventually form the nervous system, begins to form and cells to differentiate. Warnock emphasized that even if the prefertilized egg and the sperm are alive and human, the issue is when we should start treating the fertilized egg as we would another human being. In her view, ultimately that was a matter for society to decide, taking into account the scientific evidence. However, the changes taking place in the embryo at fourteen days provided a marker for the true emergence of a human individual. The existence of this criterion provided an answer to those who wished to argue that any use of a human embryo would be the start of a slippery slope to greater permissiveness.

The decision to allow the use of human embryos up to fourteen days after fertilization resulted in a considerable discrepancy from the law on abortion which, under certain circumstances, then allowed the destruction of a foetus up to twenty-eight weeks of gestational age. Whether abortion should be permitted in any circumstances has been, and still is, a matter that has aroused exceedingly strong feelings. In the USA the 'pro-life' anti-abortionists even went so far as to murder a doctor who worked in an abortion clinic. Unfortunately, it became a black and white issue, the two sides talking past each other in public debate. Warnock pointed to two relevant considerations. First, should the pregnant woman have a right to choose the course she will take, or should the foetus have an inalienable right to life? Second, a practical issue: even if abortion were made illegal, backstreet abortions would continue, often under highly undesirable conditions. But, if abortion is permitted, to what foetal age should it be allowed? Here Warnock pointed out that the later the abortion, the nearer it came to murder. At the individual level, each case should be judged on the circumstances pertaining. In the absence of a clear developmental criterion, at the public policy level abortion must be regulated in a way that is acceptable to most people. It is sad that a law requiring definite limits is necessary.

Additional problems can arise in practice. In many poorer countries where abortion is legal, many doctors have no alternative to surgery; and in countries where it is not legal abortions are carried out by backstreet abortionists. 'Morning after' pills are available in some countries, and pills have been developed for the early termination of pregnancy. The World Health Organization (WHO) would like to add the latter to its list of essential medicines, but there have been considerable delays, which are believed to be due to lobbying by a US agency. This would be in line with G. W. Bush's views on the sanctity of life. If that supposition is correct, then research, pragmatics, and ethical issues are opposed.[5]

With the rapid advances in medical science and genetics, new problems are constantly arising.[6] One is pre-implantation genetic screening, which has the potential to manage, and even to eliminate, certain diseases like cystic fibrosis. It involves fertilizing a number of eggs *in vitro*, and testing each fertilized egg to see if it carries the gene or genes for cystic fibrosis. Embryos that do would be discarded, while an embryo that did not would be put back in the mother. There are some extremists who see any interference with natural processes as sinful, arguing that even to discard an unfertilized embryo is wrong, and others who see any change to the natural condition of human beings as dangerous. A quite different objection comes from those who see attempts to eliminate a certain disability as damaging to the self-respect of those now suffering from it. It must also be added that the manipulation of embryos carries a certain risk. But since the technique could do untold good by eliminating the suffering that goes with handicap, this line of research must be seen as potentially beneficial. The cost/benefit ratios are clear in the cases of muscular dystrophy and cystic fibrosis, but may be seen differently in the case of interventions intended to discover the sex of human offspring before birth. In the UK the Human Fertilisation and Embryology Authority regulates such practices.[7]

Recombinant DNA techniques can be used to manipulate the genetic constitution of individuals—i.e. so-called 'genetic engineering'. The nucleus of one cell, usually but not invariably an unfertilized egg cell, can be replaced by the nucleus from another cell, usually a foetal cell.[8] The new cell is then caused to develop into a round ball of cells. The cells at that stage have the potentiality to develop into any cell type. If introduced into a part of the body where the cells were damaged or functioning abnormally, they would develop into the appropriate organ and the malfunctioning would be corrected. Thus if a patient suffers because mutant genes cause one cell-type to function abnormally, it may be possible to change the precursors of that cell-type so that the cells subsequently function normally.

It is important here to distinguish between therapeutic and reproductive cloning. Therapeutic cloning raises the possibility of obtaining pluri-potential 'stem cells', to cure Parkinson's disease, cystic fibrosis, and other clinical conditions.[9] Such procedures could bring great benefits to the individual patient, but would not affect subsequent generations. Apart from the question of unexpected side-effects, such a procedure would seem not to raise moral problems for most people.

The other type of procedure, reproductive cloning, aims to allow the stem cells to develop into a new individual. This was first achieved in the famous case of Dolly the sheep and has led to enormous controversy[10] because it raises the spectre of squads of identical humans. In fact that possibility is extremely remote, but it obviously poses severe moral problems. The fact that successful techniques for modifying the germ-plasm lie some time in the future means that there is time to act, and it is imperative that the possible consequences of such techniques should be considered now.

Indeed any genetic manipulation that involves changing the genetic constitution of the germ cells involves important moral decisions. If the goal is simply to eliminate a disability, as in the cases discussed above, many see the matter as uncontroversial. But the development of individuals with certain characteristics immediately raises the questions of which genes should be replaced, and what characteristics are desirable in the offspring. Suppose that such attempts were successful, they might be used with the best intentions—to improve disease resistance, or to make humans grow taller or live longer. But who is to predict what the world will be like several generations in the future? What would be the effects of longer life or taller individuals on society? And who would decide what is desirable? Would such decisions be the result of a particular political dogma? In any case, 'designing' babies is likely to be a financially demanding procedure, and the possibility has been raised that it will lead to distortion of the sex ratio, or to a division between genetically advantaged children of wealthy parents and 'natural' babies. Do such risks outweigh the possible benefits? Here is another case in which science and technology bring innovations seen as potentially beneficial by some but regarded with suspicion by others, sometimes but not always justifiably. In the UK reproductive cloning is prohibited but therapeutic cloning permitted under regulatory control. In the USA religious fundamentalists and pro-lifers who, perhaps believing that this type of research could bring harm to society, have succeeded in banning both.

As another point, some have argued that attention has focused on the tangible consequences of scientific advances, and insufficiently on the metaphysical ones, for instance on possible changes in our values, beliefs, and norms: were human reproductive cloning to be accepted, it might change our whole perspective on life.[11] But prohibiting genetic manipulation would involve prohibiting research that could relieve currently incurable diseases and alleviate much suffering.

A new development raising further ethical issues is pharmaco-genetics. Individuals vary in their response to the same treatment, and some of the variation is due to differences in their genetic make-up. The possibility of designing medicines to suit the individual raises questions about how such medicines could be tested for efficacy and safety, the consequences of labelling individuals according to their genetic constitution, and so on.[12]

That restrictions are necessary and must be enforced is shown by cases where harmful consequences could have been predicted but nothing was done. An interesting one is that of the smallpox genome. At one time the WHO was able to state that the disease had been eradicated worldwide. However, small quantities of the agent were permitted in two laboratories, one in the USA and one in Russia.[13] This has enabled the genome to be sequenced and the results published on the web. As a result, it could now be reproduced, and the disease could exist for ever. As another somewhat different example, the artificial synthesis of poliovirus by reconfiguring DNA resulted in the artificial virus being made available by mail order. Again, the insertion of a single gene into the mousepox virus, carried out for another purpose, unexpectedly enabled the virus to overcome both genetic resistance and immunization to the disease. And nanotechnologies have enormous potential for useful applications in the future, but it has already been shown that some can be dangerous to human health.

It is necessary to recognize that following a given line of research simply because it is possible may not necessarily be in the best interests of society, and ethical committees have been set up to monitor and control the ethical content of research. A number of powerful technologies currently under development, such as genetic engineering, have enormous potentialities both for benefiting and for harming human societies, and their development must be carefully controlled. Indeed any aspect of biological, medical, and psychological research may be misused, and in the UK and many other Western countries all research is properly subject to ethical approval by responsible committees. Every research project that involves human or animal subjects has to be approved by an ethical committee of the university, hospital, and/or grant-giving body, and also by the journal to which the work is submitted for publication. It is the duty of the ethical committees to ensure that the investigation does not put the subject's health or welfare at a significant risk or be likely to produce consequences harmful for the community. In the case of animal experiments, the work must not cause suffering unjustifiably by its predicted benefits. It still remains the case that much chemical and biochemical research that does not use human or animal subjects is unrestricted.

It is also desirable that such committees should include lay members. As we have seen, many of the problems arise from clashes between world views. While the expert knowledge lies with the medics, lay representation helps to ensure that decisions are not taken solely by medics seeing themselves as a group with superior knowledge and a duty to their profession.

A continuing need is for standardization of the criteria used between committees. Although regulation is essential, it seems difficult to devise procedures for doing so that do not involve a great deal of bureaucracy which could hinder the course of research.

PREMATURE PUBLICATION

Another temptation to which scientists in all disciplines are subject involves the publication of results that are based on insufficient data or erroneous interpretation, especially when publication may have an impact on society. Here the problem is likely to be the career ambitions of the research worker in conflict with his scientific integrity. Usually such publication is not in a scientific journal, where it would have had to pass peer review, but directly in a release to the media or the internet. The public generally tends to believe in sensational findings, and improper publication may lead to a great deal of anxiety or suffering.

The worst case is when the scientist concerned persists with his findings, even when they have already been shown to be false in further, more thorough, investigations. A recent case in Britain concerned the advisability of vaccinating children for measles, mumps, and rubella in a single injection instead of separately. One research worker claimed that simultaneous administration was dangerous, causing harmful effects on the children: many parents therefore refused to have their children vaccinated. This resulted in a widespread disruption of the Government's vaccination programme which was aimed at reducing the incidence of these diseases in the population. The data were subsequently shown to be incorrect, though the scientist persisted in promoting them for some time.

The obverse of premature publication is failure to publish negative results. It can happen that successive drug or treatment trials yield contradictory results. Tracing the bases of the discrepancy in subjects or procedure can yield important new insights.

ANIMAL EXPERIMENTATION

Experiments on animal subjects are essential in medical and biological research.[14] The use of quite large numbers of animals is legally required for testing new drugs before clinical trials with human subjects begin. Smaller numbers are used in basic research and to facilitate the understanding of many human diseases: practically every form of medical treatment has depended on research with animals. Regrettably, however, the ethical problems involved in the use of animals have been adequately faced only in the last half-century. In many Western countries there has been intense public and philosophical debate on if and when animals should be used to further scientific knowledge and to test medicines and medical procedures. Some take the view that animals should never be used in research, others that the probable benefits to human and animal welfare must be weighed against the suffering caused. Patrick Bateson, in an extensive review of the debate from both scientific and philosophical perspectives,[15] suggests that three separate dimensions should be assessed independently, the scientific quality of the research, the probability of human benefit, and the likelihood of the animal suffering. (The controversial inclusion of the dimension of research quality would permit research of high quality and involving minimal suffering even if of little predictable benefit to humans.) The problems involved are complex, and involve assessing human benefit against the suffering caused, neither being susceptible to precise measurement—though indirect methods for assessing animal suffering are available. In the UK the result of the debate has been the introduction by the Home Office of rigid procedures to regulate experiments on animals.[16] These are reinforced by local ethical committees and by the refusal of journal editors to publish research that involves animals in unnecessary suffering.

In Chapter 2 it was argued that the human propensity for prosociality was directed primarily to members of the same group. It is instructive to note how the debate about the use of animals in research has been influenced, perhaps unconsciously, by the extent to which animals are seen as related to humans. Two criteria have been used. First, the use of domesticated species, cats, dogs, and horses, seen perhaps as part of the human household, has been subjected to much more rigid restriction than the use of any other species. More recently, greater emphasis has been placed on complexity of cognitive development and the associated supposedly greater similarity to humans in the capacity for suffering. The use of apes, which are cognitively and biologically closer to man than any other species, is forbidden in the UK and the use of monkeys is severely restricted. The restrictions on the use of Old World monkeys are greater than those on the supposedly less advanced New World monkeys, though this is probably due to an underestimation of the cognitive capacities of some genera of New World monkeys.[17]

At present efforts are being made to reduce or improve the use of animals in scientific research in three ways. First, the number of animals used in each test or experiment should be reduced to the minimum necessary to obtain meaningful and valid results. For instance, treatment of an experimentally induced condition formerly required the sacrifice of groups of animals at successive time points so that the effect on the condition could be assessed post mortem. Now it is sometimes possible to follow the course of the treatment by the use of magnetic resonance imaging. The same animal can be scanned a number of times and changes in the internal organs monitored. Second, experimental techniques are being refined to avoid or minimize pain and discomfort, and to obtain the greatest possible amount of information from each test. Third, alternative techniques, involving for instance tissue culture, are being used to obviate the need for using animals altogether.[18]

MEDICAL PRACTICE

Trainee doctors must gain experience with patients, but special precautions must be taken in the use of patients for teaching. For the good of the patients for whom students will later be responsible, it is essential that they be given as much first-hand contact with patients as possible, but there are always dangers of infringing the patient's dignity even when consent has been obtained. Students' training should include discussion of the ethical issues with which they will be confronted, as well as the purely medical ones.

For example, the doctor's view of what is best for his patient may conflict with the ethical principles of his patients. While the doctor almost certainly knows more about the treatment than the patient he is trying to help, his knowledge may be held to be irrelevant by the patient. Some Jehovah's Witnesses believe that it is wrong to receive a blood transfusion: this may conflict with the doctor's duty to do his or her best to save the patient. Even more extreme cases may arise when a patient is on a life-support machine. Should the decision to turn off the machine be governed by Hippocratic principles, the patient's previously stated wishes, or those of relatives?

Fortunately, malpractice is rare. Two thousand years ago, when the Hippocratic oath[19] was introduced, and still today, the life of a patient was literally in the hands of the doctor, and it was essential to ensure that the doctor would wield his power responsibly, with the care of the patient being his foremost duty. But specific guidelines are now necessary, and many medical procedures are regulated, often by government-appointed bodies. In the UK the General Medical Council in England has the power to stop a doctor found guilty of professional misconduct from practising by withdrawing his licence.

During the twentieth century, the relations between doctors and patients changed dramatically in a number of ways, of which I shall mention two. These changes have been due partly to the greater medical knowledge that the public believes itself to have and partly to patients' greater unwillingness to accept authority. First, early in the twentieth century (and I write in broad generalizations with the approximations that that entails) many doctors felt it their duty to reassure their patients, which often meant understating the seriousness of their condition. This may have been good practice, for the health benefits of optimism are well known. Only if death was imminent or inevitable would most doctors break the news. This also may have been wise in a more religious age, for it is more comforting to look forward to the Everlasting Arms than to be told that you have a one in twenty chance of survival. Nowadays doctors are expected to tell the truth, but it is necessary for them to exercise great sensitivity in this respect. Some patients may not be ready for the diagnosis, and denial can be an important defence mechanism. What has changed is an increased demand by the patient for the truth: indeed, in the UK everyone now has the legal right to see his or her medical records. This also has its merits in creating a more open relationship between doctor and patient and encouraging the patient to trust the physician.

A second way in which doctor/patient relations have changed is that, while formerly doctors told their patients what the proper course of treatment should be, now there is an increasing tendency for doctors, when alternatives are available, to allow the patient to chose from a range of options. Furthermore, doctors must ask for the patient's consent or, when that is not possible, for that of a relative, before many procedures are carried out. Of course, in some cases involving accidents or acute conditions the patient's consent cannot be obtained, but in most others it is mandatory. These changes would appear to remove some responsibility from the doctor. However, the doctor is likely to have his own view as to what is best, and the patient's choice is likely to depend on how the doctor explains the various options.

Certainly it is incumbent on the doctor to inform the subject of possible side-effects. But the doctor may explain some options more fully, or in more rosy terms, than others, and thereby influence the patient's choice: patient consent then becomes something of an illusion. In any case, except in matters of urgency, the patient should be given time to think the matter over. And should the doctor's advice be influenced by an estimate of the patient's likely reaction or ability to understand? The increasing knowledge possessed by patients in the twenty-first century, enhanced in recent decades by the internet, is an important issue to be taken into account in considering the ethically preferable course for the physician.[20] And if the doctor takes the issue of patient consent seriously, how is he to be sure that the patient, who lacks his knowledge, has understood the relative risks fully? Consent has only the appearance of absolving the doctor from responsibility.

Related to these matters is the possibility for legal action. Legal cases involving the prosecution of doctors for improper or inadequate advice or treatment have been commonplace in some countries for many years, and are becoming common in the UK. According to an official of the Medical Defence Union, fear of litigation is affecting clinical practice and morale.[21]

Influencing all these changes is the question of medical fees. Before World War II, a few 'panel' patients obtained medical treatment free or for a very small weekly fee in the UK. However, 'private' patients were charged a fee by the doctor. The size of that fee was a matter left to the doctor, who would be likely to charge what he thought the patient could reasonably afford. While this gave doctors who were so inclined the opportunity to behave altruistically towards their poorer patients, the majority inevitably preferred patients who could afford to pay, and were likely to spend more time with the rich than with the poor. And the system inevitably meant that there were many ill people who could not, or were unwilling to, afford medical treatment. Now, in theory at least, medical care in the UK is a matter of right for everybody, with all having equal access to the best care. However, some private practice continues, and with it continuing controversy. Some feel that the National Health Service is undermined by the existence of private practice and private medical insurance, for it is still the case that the rich can buy better care than the poor. Others argue that private insurance is relieving the state of some responsibilities, thereby making more available for those who cannot afford to pay.

From the physician's point of view, the changes may be seen somewhat differently. In the days of private practice, while some exploited the rich—easy enough when life itself was at stake— the majority of doctors felt impelled by their Hippocratic oath to do their best for the patient. For many such doctors the pressure of felt responsibility was immense: night calls were never refused and time for relaxation was bought at the cost of a heavy conscience.

ORGANIZATION OF MEDICAL SERVICES

In addition to the issues facing the practitioner of medicine, even more far-reaching problems confront those who organize medical services. Marginalized groups in societies are especially vulnerable to health problems: what can be done to help them in a democratic society? Should they get a greater share of medical resources? Another issue concerns the conflict between individuals' rights to freedom of movement and the need to segregate individuals or groups to prevent the spread of communicable disease. And how should the WHO distribute its inevitably inadequate resources to maximum effect? These and many other problems are ultimately questions of human rights.[22]

CONCLUSION

In this chapter, I have deliberately stressed the nature and diversity of the ethical problems that arise in medicine. Many of these have long been with us, and the medical tradition of responsibility to patient and community is strong. As in the physical sciences, both medical research workers and practising doctors are proud of their profession and feel a strong sense of duty to maintain professional standards. The medical researcher or practitioner is seldom forced by the situation to act in ways contrary to his basic code but, as a safeguard, in the UK and most other Western countries, appropriate mechanisms for controlling both medical research (ethical committees) and medical practice (the General Medical Council) exist.

It is, therefore, probably inappropriate for an outsider to point to the medic's Achilles heel. Nevertheless it is an issue that has already been mentioned in the previous chapter and will appear in later ones. The medical researcher knows how important the clinical trial is, and the medical practitioner knows he is right to want to do his best for the patient, even though the patient may hold views that seem to him misguided. Even when it is a matter of the rights of the individual versus benefits to the community, the medic has superior knowledge and can feel that he is taking an objective outside view. Long training may not only assure him of his own judgement, but may actively encourage him to display his certainty. In addition, he sees himself as upholding the ethical precepts of the caring profession to which he belongs. And there, like a small cloud on the horizon, is the danger. It is not difficult for him to convince himself of his own rectitude.

7

Ethics and Politics

Although politicians are only human, they must take part in momentous decisions affecting the lives of millions. They may even have to decide whether or not their country should go to war, with the inevitable suffering that that will mean both to their own countrymen and women and to the perceived enemy. Most of the decisions they take must be made with inadequate evidence. Dilemmas involving both practical issues and moral principles are intrinsic to the life of most politicians. They raise the question of how politicians justify to themselves the decisions they take.

Inevitably, opinions about political matters vary, and politicians come in for a great deal of criticism. The aim of this chapter is to specify some of the conflicts to which they are exposed, and to ask how much the criticism they receive is the result of the system in which they operate, how much to the application of an amended moral code ('political morality'), and how much to their own failings. While this book is aimed at general issues, I have found it impossible not to be influenced by recent events, and have used my personal interpretation of recent political decisions to illustrate some of the points discussed.

MORAL CONFLICT INTRINSIC TO THE POLITICAL SYSTEM

How political leaders come to occupy the positions that they occupy has an important relation to their attitudes to moral problems.

Consider, first, a totalitarian system. One must allow for the possibility that its leader has reached his (or her) position because he is seen as genuinely concerned for the welfare of the people. Certainly some seem to have been benign: Joseph II of Austria and Frederick II of Prussia have been called 'enlightened despots'.[1] But it is more likely that he has got to where he is primarily because he is ambitious and unscrupulous. In that case, if he is aware of moral principles, they will have negligible effect. His position is based on selfish assertiveness, and he will give priority to securing his power and furthering his own material well-being. Whether or not he perceives ethical conflicts, they have little influence on the decisions he takes.

But is a democratic system always beyond reproach? We can forget the cases of rigged elections, or the use of coercion, as they do not count as democratic elections. In an ideal democratic system the leader is elected because he is trusted by the electorate and/or by a group of elected colleagues and is seen by them to be the person most suited for the post. But a number of problems are intrinsic to the democratic process.

A basic issue is that more than one electoral system is possible, and it is arguable which most fairly represents the electorate. This is not the place to enter the lists over the advantages and disadvantages of different electoral systems, though many feel that proportional representation would produce a more representative outcome than the system at present in place in the UK. Sadly, any change in the electoral system will probably occur only if the party in power sees that that would be to their advantage. Again, some elections are based on the prior results of the election of regional representatives and, although they may have to vote according to the majority in their constituency, that may result in the election of an individual who would not be the preferred candidate of a majority of the whole population: that seems to have been the case in the 2003 presidential election in the USA. And if the leader is the personal choice of a number of colleagues who have themselves been individually elected, how far is their support governed by their own longer-term self-interest?

Second, in seeking election a politician must tailor his own views to some extent to the inclinations of the electorate. Thus when John Kennedy was up for a second term in the Senate, he downplayed his growing disapproval of the McCarthy anti-communist witch-hunt because he believed a high proportion of the Massachusetts electorate to support McCarthyism—a course which he subsequently regretted.[2] Presumably he justified compromising his integrity as necessary for a greater goal, or perhaps allowed his ambition to blind him to the issue. Furthermore, a candidate may direct his election campaign to sections of the community whose votes he sees as most important for his election—to black voters, or trade union members, or middle-class voters as seems appropriate to him. For instance, in the UK the disregard of politicians for those who live outside towns has been explained by the fact that their small numbers account for too small a percentage of the votes.

More importantly, ideally the electorate should base their votes on their personal estimates of the candidate's relevant abilities and the policies of the party he represents. In practice most electors in Western countries have to rely on the media, especially television and newspapers. And in some countries the coverage candidates or political parties receive depends largely on what they can pay for, which in turn may depend on how many of their supporters are very rich and the size of the contributions they receive from big business. It may even be the case that contributions to a political party are made in the expectation of advantages subsequently to be bestowed: a culture based on material wealth is inimical to true democracy. Although there is now a ceiling on the money that a political party can spend on the election process in the UK and some other countries, it can still be the case that some parties' funds are inadequate to reach that level, so that they can buy less advertising time and space than their competitors. We accept such problems as part of the democratic system, but we should be aware that democratic systems, though potentially better than any other, do not inevitably lead to a fair representation of the merits of the candidates or the opinions of the electorate.

Party politics can form another obstacle against the selection of the best candidates. Candidates for high office may be selected by the party bureaucracy, who may favour candidates who have shown long-term loyalty to the party. Loyalty is a virtue, but that does not mean that it necessarily characterizes those who are best fitted for governing. Or the party bureaucracy may choose members of one gender or race on the grounds of party expediency.

In the UK Parliament, and also in the US Senate and House of Representatives, party politics sometimes constrains how members vote in debates. The 'whip', requiring them to vote with the leaders of their party, places many in a conflict between loyalty to the party, their own ambitions, and their own integrity. The party machinery has helped in their election, and loyalty to the party is seen as a moral ought which must override personal beliefs. If they do not vote as the party demands, they may lose any governmental office they hold, reduce their chances of preferment, or even suffer exclusion from the party. If they follow the party line, they may further their chances of promotion. Thus loyalty is reinforced by carrot and stick. This was probably one of the issues that enabled Blair to obtain parliamentary support for the second invasion of Iraq. The issues here resemble those that arose in the evolution of prosociality: personal assertiveness must be restrained for the sake of the well-being and competitive success of the group as a whole. The concept of loyalty has arisen as a way of meeting this dilemma.

To be fair, it is difficult to devise a system that would overcome all these problems, so let us bypass such worries and consider how a democratic leader is likely to be placed in moral conflicts by the very fact that he has been elected. Political decisions are always likely to involve conflict between incompatible desiderata, and moral issues will be prominent. There are difficulties in weighing up the various consequences of any action—the goal and the probability of achieving it, the views of colleagues and of the Opposition, the responses of third parties, the inadequacy of the information to hand and the effects on the electorate. Often a leader may be taking decisions under great stress.[3]

Situations must arise when the politician's convictions as to the right course to take are contrary to those of the majority of the electorate. Which course is the 'right' course must then be a matter for heart-searching. When the two are not in accord, there may be good reasons why sometimes his convictions and sometimes the wishes or beliefs of the electorate should be given the more weight. But there is a real danger that leaders may be unable to get past their convictions that they are right. It certainly appeared that the Second Gulf War was launched in part because the two main leaders in the coalition failed in exactly that way. Bush argued that a leader's job is to lead the electorate, and not to be led by it. This was the line he took when, in trying to convince the Italian prime minister to join the coalition before the Second Gulf War, he argued that 'We lead our publics. We cannot follow our publics'.[4] Blair showed the same determination not to budge as did Bush.[5] It is reported that, having clandestinely agreed with Bush that Saddam Hussein must be overthrown by military means, Blair declined to pull out of the coalition when Bush gave him an opportunity to do so, even though he knew that there was a real danger that his Government would fall as a result.[6] He must also have known that war would have horrendous consequences. If politicians can disregard the opinion of the electorate on the basis of their own conviction, there is a great danger that they will become obsessed by perceptions of their own rectitude, and act as if they were dictators.

Alternatively, a leader may feel that it is his duty to represent the views of the electorate. Then the question arises, how much is this due to a desire to keep the electorate on his side so as to improve his chances of re-election next time? And if he feels strongly that the electorate is wrong, should he not resign?

One may also ask how well he knows the views of the electorate. As we have seen (pp. 25–7), it is easy to ignore contrary views, or to devalue the sources of contrary opinions. Amartya Sen has emphasized that democracy involves not only elections but 'public reasoning', including the opportunity for public discussion.[7] How much that desideratum is met in the ordinary course of British politics is extremely dubious.

Finally, the electorate is not the only source of pressure on politicians. If a leader is to persuade his colleagues in government, and perhaps also those in opposition, to support the introduction of legislation that he sees desirable, further compromises and wheeling and dealing may be necessary.

There is no suggestion here that democracy is not, in most circumstances, the best system: my aim has been to show that politicians in a democracy must face many difficult ethical conflicts. And for politicians at every level, party loyalty, a form of group loyalty based in a long history of biological and cultural evolution, has come to be seen as a moral issue.

MORALITY IN POLITICAL DEALING

This is a difficult issue to discuss in general terms, so I shall again use the ethical aspects of the events leading up to the Second Gulf War to illustrate the problems with which political leaders may be faced. Blair did all he could to persuade Parliament that he was right by arguing that there was a direct and imminent threat to this country and the world from the possession by Saddam Hussein of huge arsenals of chemical and biological weapons, and possibly even nuclear weapons. The prime minister based his argument on intelligence reports when he must have known that the 'facts' were probabilistic. Subsequently it became clear that this dossier was based on unreliable intelligence information and the presentation may have been selectively emphasized by his own staff. The question arises, did he allow his judgement to be affected by his earlier conviction that war was the right course?

It is idle to speculate whether the House of Commons would have approved going to war in the absence of a dossier issued with the prime minister's authority. The probability is that it would not. But, be that as it may, it subsequently became clear that the threat had been virtually non-existent. Despite a long, intensive, and costly search by a very large team of scientists and technicians appointed by the US Government, no evidence of ready-to-use weapons of mass destruction could be found. It is now clear that Blair misled Parliament and the British public. It may not have been a deliberate lie, but it involved the blithe acceptance of the intelligence report. In doing so, Blair took a decision which led to a war seen by many as illegal and to unmeasurable human suffering, and was probably partly responsible for subsequent terrorist attacks in the UK and elsewhere. If politicians were guided by ethical considerations, he would have resigned by now, or, at least, apologized to Parliament and asked for forgiveness. In old-fashioned terms, that would have been the honourable thing to do. But he did not, and had no intention of doing so. He shrugged it off, and let ethics be damned.

In a speech delivered in March 2004 Blair argued that he had long been concerned with the threat of global terrorism. Regime change alone had not been a reason for going to war; the primary reason was to enforce UN resolutions over Iraq and WMD. 'We had to force conformity with international obligations that for years had been breached with the world turning a blind eye'.[8] The hypocrisy of this statement when the UK and other nuclear powers were failing to meet their 'unequivocal undertaking . . . to accomplish the total elimination of their arsenal leading to nuclear disarmament . . . '[9] is not the present issue. The point is that he saw it as necessary to 'take a stand before terrorism and weapons of mass destruction come together, and I regard them as two sides of the same coin'.[10] In fact, there has been no evidence that Iraq was a stronghold of the terrorist group Al Qaeda, which in fact was then based in Afghanistan, and Iraq did not harbour weapons of mass destruction. That it might have done in the future is beside the point.

Blair must be held responsible for the consequences of his decision, but why did he take it? It looks very much as if he was carried along by the inertia of personal conviction, perhaps exacerbated by his religious beliefs. Here the terrible demon of moral rectitude has raised its head. It is too easy to think that we must be right, for we are equipped with mechanisms predisposed to reassure us of our own rectitude by maintaining congruence between how we see ourselves to be behaving and the ethical precepts in our self-systems. Because, in a democratic system, the prospect of re-election must always weigh with politicians, they are especially likely to find it difficult to admit that they have been wrong—even more difficult if they believe they have received divine guidance, or have been furthering a righteous cause.

Even if one makes the most charitable interpretation of Bush's and Blair's motivation to attack Iraq, nothing can excuse their attempts to find retrospective justification for going to war. Lacking evidence for weapons of mass destruction, both Blair and Bush claimed that they had gone to war to unseat a tyrannical regime, or as part of the so-called 'war on terror'. Regime change was certainly discussed before the war, though it was not presented as a major justification to the public. It may have been a consequence of the war, but whether there has been a decrease in suffering is still, at the time of writing, highly questionable.

The run-up to the Second Gulf War provides many other examples of actions that many would consider morally dubious. To give a few examples, how far should a politician, who is following the course that he thinks is right, go in deceiving not only his own electorate but also foreign governments in order the better to achieve his goals? According to the account given by Woodward,[11] Bush, by playing with words, effectively lied to the German chancellor and the French president in saying 'I have no war plans on my desk' when extensive planning was underway. Powell, his Secretary of State, tried to 'ambush' the French, who were opposed to war, with a supposed ambiguity in the wording of a Resolution for the UN justifying the war. American intelligence spied on the activities of the UN weapons inspector, Hans Blix, who was trying to discover if Iraq had weapons of mass destruction.

Again, how far should a democratic leader attempt to 'lead' the electorate, attempting to persuade them to support a plan already laid out. The US president asked the Intelligence Services to put together the 'best information' favouring war, and lawyers to use it to make the best possible case. It is clear that Cheney and other members of the Bush entourage interpreted the evidence according to their own pre-existing biases, those favouring war converting uncertainty and ambiguity into fact.

A related issue concerns the use of rhetoric and euphemisms in political dialogue. Almost inevitably, it seems, political leaders must arrange for the information that goes out to the public to be sifted, and exhort their public by calling on concepts of loyalty, courage and determination. In time of war they use euphemisms to conceal the war's horrors. There is a fine line between persuasion and psychological coercion.

One cannot be sure that such moves involved deliberate deception. It is so easy to be guided by the illusion that one is right. In any case, many see such actions as an inevitable part of the political game. They will argue that, like chalk and cheese, ethics and politics simply don't go together: deception is one of the tools of the politician's trade. While everyday moral principles have been elaborated in order to smooth relations between individuals in small groups, politicians must make decisions over much more far-reaching issues, so that 'political ethics' are different. Or is that just a kind way of explaining away the actions of a minority of politicians who make it inevitable that politicians should be seen as the least trustworthy of professionals? Are the actions of this minority straightforwardly the consequence of something like personal ambition driven by selfish assertiveness and justified post hoc by moral rectitude? Or are they simply and genuinely misguided as seen from one's own point of view? Certainly, ascribing their actions to 'political ethics' may be valid for outsiders, but it is doubtful how often politicians say to themselves 'This is a political decision and not subject to everyday moral principles' in the way that a businessman may justify his actions by 'business ethics' (see Chapter 8). Do they tell themselves that the end justifies the means? My guess is that many think 'I have just got to do this, and that is all there is to it'. Or should we point the finger at the situation: some double-dealing is inevitable because a politician is likely to be caught between the conflicting spurs of his own political views and the desire to retain office?

Nothing in the above implies that all politicians are untrustworthy. Many, possibly most, politicians are genuinely concerned with the well-being of their electorate and indeed of the world. In the case of the second Iraq war, one honourable politician put his career in jeopardy by resigning from the UK Cabinet when he could no longer honestly support the collective view. Another followed later. We must hope for more like that, and for greater honesty and openness in political dealings.

ASSESSING POLITICAL DECISIONS

Given the difficulties involved in many political decisions, how should one judge the decisions taken? In everyday life we may excuse a wrongdoer because we recognize that his actions were guided by values different from our own, or because he 'meant right', or because he was ignorant of the consequences. Do such excuses apply to politicians?

The course a leader takes must depend on his motivations and values. Suppose, to take an extreme example for the sake of discussion, that Stalin was not a megalomaniac and genuinely believed that Communism and the security of the motherland must have top priority. He is said to have been a kind man in his family life. Suppose he really did wrestle with the conflict between the means he found to be necessary to achieve his ends, and the consequences of those ends: would that have justified the horrors for which he was responsible? If one has to make a judgement, should that be based on his actions, his values, or on how he came to his decisions?

So far as his actions go, they clearly did not involve prosocial reciprocity. No one could argue that it was right to put so many to death, to imprison or exile countless others, to impose a way of life on the citizens, in the perhaps vague hope that it would lead to a brighter future for those who survived to see it. Admittedly we come from a moral climate in which individuals are valued: leaders from other cultures may place greater value on the collective. And it may be easier for us to say 'it cannot be right' because we are attuned to thinking in the moderately short term, not in terms of future generations. But such considerations are surely insufficient to make Stalin's actions acceptable.

What about the values that were guiding him? Suppose he did genuinely believe that he was doing the right thing, or at least the best that he could. If one sees the values in isolation from their possible or actual consequences, can one then understand them? Surely that would be giving Stalin too much credit: his actions were too clearly contrary to the Golden Rule of Do-as-you-would-be-done-by. One cannot condone hideous actions because the actor thought he was right: the road to hell is paved with good intentions. If one justifies tyranny on such grounds, one must condone also the Holocaust, the Inquisition, the excesses of the French Revolution, and every self-righteous individual who thinks that the means are justified by the ends.

But there is another issue: suppose he did not know what consequences his actions would have. In the everyday world, if a well-intentioned act has unfortunate consequences, we do not consider that the actor has behaved immorally. Not, that is, unless he should have known what the consequences would be. That raises the question of how to judge whether an actor *should have* known. In fact there is a continuum from say, giving a child peanuts when it turned out that the child was allergic to them, to a doctor who prescribes a medicine without reading the counter-indications on the label. In the former case the donor might not have known that the child was allergic to peanuts. After all, most children are not. But the potential dangers of the medicine *ought* to have been known to the doctor. Where one draws the line between what the actor might not have known and what he should have known is yet another issue. But in the case of a politician taking a decision that will affect the lives of many, it is surely incumbent on him to do all that is humanly possible to investigate the possible consequences. Ignorance is no excuse. And in this case, Stalin knew the consequences: perhaps he did not really reflect on what they entailed.

One must think also about how such decisions are made. Perhaps he was brought up with skewed values, so it was not really his fault—his parents, or his environment, or the bad lot he fell in with in his teens. Or perhaps he was endowed with too large a dose of certainty in the validity of his own opinions—moral rectitude again. Such excuses may be relevant to individual wrongdoers, but they provide too easy a cop-out for those who make decisions with such a widespread impact. The leaders in cases such as those with which we are concerned will have had ample opportunity to 'know better'.

Thus, morally as well as legally, it is by their actions that leaders must be judged. Good intentions do not suffice.

Finally, we may consider some possible obstacles to decision-making that stem from pan-cultural psychological characteristics.

THE EXTENT OF THE LOCAL GROUP

Here we must discuss separately internal politics from decisions concerned with external (i.e. foreign) affairs. Internal politics are often concerned with the welfare of groups within the society—the young, the aged, the disabled, trade union members, and so on. The politician is usually outside the group in question, and his task is to allocate resources 'fairly', taking into account also the needs of the rest of the population. Leaving aside the politician's almost inevitable party bias, considerable ethical problems arise. What do we mean by 'fair'? Is it according to need? Then how do you compare the need for wheelchair access to public buildings with the costs to the community of providing it? Another way to judge fairness is by the contribution of the group in question to society: but how then do you compare, say, firemen with engine drivers or with nurses? Or should one judge by scarcity value, the politician recognizing the need to raise the pay of a certain group to attract more into the occupation? These issues are difficult but commonplace and well recognized. And in all of them the politician stands (in theory at least) outside the situation. But does he? If he represents the government and is negotiating with a trade union, for instance, he will be in a bargaining position, attempting to get the best deal for the government that he can. Alternatively, he may want trade union support for his party. He may then feel justified in basing his actions on 'business ethics' (see Chapter 8).

But in foreign policy the situation is different. The politician is an in-group member. He sees it as his duty to do his best for his own country. In Chapters 1 and 2 we saw that it is likely that prosociality and in-group cooperation arose in part from inter-group competition: groups with a high proportion of individuals who behave cooperatively with other in-group members out-competing groups with a small proportion, with resultant ben-efits to the individuals in the former. Thus prosociality is directed primarily to in-group members, not to outsiders. In acting to further the interests of his own group, the politician is acting according to his natural inclinations. And if it is a matter of war with another country, the politician may feel it is his duty to denigrate the other country and to exaggerate the danger that they pose.

But it is easy for politicians to go too far. Many feel that the US administration has committed that error many times in recent years. It has disregarded international treaties and the decisions of the UN that might be construed as against American interests because the administration believed themselves to be right or their interests to be paramount.

In any case, the world is changing. Globalization has both good and bad aspects, but its progress seems inevitable. It blurs the in-group/out-group distinction because it involves increasing interdependence between the countries of the world. This makes it essential for us to think in global terms as both an ethical and a practical matter. A stock market crash in Japan reverberates round the world. War in the Middle East or the closure of oil wells in Russia affects oil supplies universally. A famine in the Sudan may have repercussions in Europe. While the stock market crash and the war may be thought of as practical issues, action to relieve the famine is surely a moral imperative.

Fortunately, as we have noted already, over the last two centuries attitudes in the Western world have shown signs of change. We now show some responsibility for victims of disasters in distant countries. But the interdependence of nations poses many questions to the politician. For instance, how much of the national budget should be devoted to overseas aid? This is a moral question as well as a pragmatic one, because prosociality to others must be weighed against the politician's responsibility to look after the electorate—though no doubt he will also be influenced by the response of the latter to his possible re-election if he allocates more than they think is just. The moral issue is, or should be, also independent from the fact that the common good may be best for the donor state in the long run.

LONG-TERM ISSUES IN POLITICAL DECISIONS

Advances in science and technology, while bringing a higher standard of living to many, are too easily used for the advantage of the few. The consequences of these advances have grave implications for the future: climate change, the excessive exploitation of fossil fuels, the felling of forests, the reduction in biodiversity, the depletion of fish stocks, the desecration of coral reefs—the list is endless. It is not my purpose here to present the case for the reversal of such changes for the sake of future generations, but consideration of the bases of morality takes us to the nub of the issue.

In Chapter 4 we discussed the role of reciprocity in relations between individuals. One individual behaves prosocially to another in the expectation that he will receive fair recompense in due course. Trust that the recipient of the prosociality will not renege on his obligation to repay is required. On the whole we find it easier to trust if the delay is likely to be short, and the return fairly certain. It is this that makes it difficult for politicians to legislate for the long-term future. The electorate is more interested in the here and now, and has less interest in the long-term future of humankind. We expect quick and certain returns, and are less willing to accept laws that curtail our current well-being in order to safeguard the possible welfare of future generations. They will not repay their debt to us. And if the voter has never been to a rain forest, and cares little about it, he is not likely to see it as important to his descendants. The legislator is thus faced with a dilemma: he may see protecting the future generations to be the morally right course, but risk the wrath of the electorate if he pursues it. But there is one thing that may help. We are so constituted that we strive to pass on our genes to our descendants. Perhaps if the matter is presented to the electorate in terms of their children and grandchildren rather than humankind or the 'future', they will be more willing to forgo short-term gains.

THE INFLUENCE OF THE PAST

Motivation for revenge has played an important part in the relations between individuals, families and groups throughout history. Even in some parts of Europe blood feuds are still seen as the norm. However, the politician should look to the current and future good of his constituents, and must not be too influenced by the past, for the past cannot be changed. Unfortunately, politicians do not always act that way. Past injustice seems to require recompense, and revenge as a powerful human motive operates in the collective memory of groups just as in individuals. The long-standing antipathies between Greek and Turk on the divided island of Cyprus,[12] between Catholic and Protestant in Northern Ireland,[13] and between Hutu and Tutsi in Rwanda,[14] go back hundreds of years. Leaders in Northern Ireland and Cyprus, acting on behalf of their own group (and perhaps to further their own interests) have kept the antipathy alive by encouraging different perceptions of history. In the case of Rwanda, the Belgian administrators helped to perpetuate the division between Hutu and Tutsi by issuing distinct identity cards, perhaps on the divide and rule principle. If such breaches are to be healed so that communities can move forward, perceived past injustices must be forgiven and forgotten. The wise politician may see this when the population does not because it perceives any attempt to give ground to the other side as treacherous. But there are some wonderful examples that should be followed. Mandela, who suffered imprisonment and hard labour for many years, showed that the only way forward was to forgive the past. And, through the example of such men as José Ramos Horta, East Timor was able to build a new country by moving beyond the injustices suffered at the hands of Indonesia.

The healing of a breach may require much more than a ceasefire, peace accord, or treaty. Truth and Reconciliation Commissions in South Africa and elsewhere have gone a long way towards achieving real understanding with the goal of admission of guilt by many of the perpetrators and forgiveness by the victims.

THE PROMOTION OF DEMOCRACY

The Iraq wars have presented yet another problem to the US and UK politicians—the introduction of democracy in an occupied country. Because, in the proven absence of weapons of mass destruction, the politicians had given the replacement of an oppressive regime by a democracy as their (illegitimate) reason for going to war, they then had to try to implement it.

Democracy is desirable because it brings greater freedom and good governance to its citizens, and is associated with greater overall welfare. It has the potential to ensure that the benefits of prosperity are distributed equitably amongst citizens. It emphasizes peaceful norms of behaviour, with conflicting views unlikely to lead to violence, and a functioning legal system. The people can choose their government and change it if they so wish, and the military is under civilian control.

At present, around 60 per cent of the world's population live in a democracy. But democracy exists in many degrees. It is easy to forget that, at the beginning of the twentieth century, the UK denied votes to women and the USA to black Americans. Today a number of states, such as Zimbabwe, claim democracy but permit only token political competition.

Experience shows that the introduction of democracy to a previously non-democratic state is a matter of great difficulty, and especially in a country devastated by war. Democracy requires an adequate infrastructure, involving not only basic needs like schools, hospitals, energy, and transport but, perhaps more importantly, trained administrators capable of building and maintaining it. Corruption must be eliminated, and that requires an adequate legal system and the means to enforce it. There must be a free press, and an educated opposition, for the voters must understand the issues. Voting must be accessible and secret. Demobilized soldiers must accept civilian control.

Many national borders were established by colonial powers without regard to tribal or ethnic differences, and voters may face a dilemma between religious or ethnic loyalties and social desiderata. Where people are likely to vote along ethnic or religious lines and not according to the issues or the acceptability of the candidates, some form of institutional power sharing is essential to prevent dominance by the majority party.

Such issues, the fruits of experience in other countries, shows that democracy cannot be imposed and must be introduced gradually. At the time of writing it is clear that the difficulties had not been sufficiently recognized by the occupying powers in Iraq. In a rush to fulfil their rash promises to their electorates, they have underestimated the difficulties. They are faced with a dilemma that seems insoluble, or at least will take years to solve. The country is beset by violence. Islamic groups, differing in their religious traditions, compete for power. Some of those responsible for the violence, belonging to the group that formerly led the country, see that democracy will bring an end to their dominating position. Many hate the occupying powers for their interference and seek to destroy law and order and the country's infrastructure: to this end they attack all who would help the establishment of a functioning infrastructure. Others from outside the country have a similar aim.

CONCLUSION

At the beginning of this chapter we asked how much of the behaviour of politicians that seems questionable is due to the political system itself, how much to a special form of 'political ethics', and how much to their own failings. In at least some of the cases discussed, their own failings seemed to be prominent, manifest especially in moral rectitude and an inability to admit error. How the politician balances his convictions against the views of the electorate requires an ability to admit his own fallibility, and, partly because of the electoral system, it is difficult for politicians to admit misjudgement.

To a considerable extent, the politician is forced to behave in that way by the system: in a democracy it is almost impossible for a politician not to be influenced by his own ambitions, and he must convince others that he is worthy of office. Of course his ambitions may be worthy ones, for he may feel that he has a message to convey—though, as we have seen, that in itself carries the danger of moral rectitude. Sometimes he may feel that the ends justify the means. Maintaining congruency in his self-system must provide many opportunities for self-deception. Compromise, often itself a virtue, is intrinsic to the democratic system. It is issues such as these that too easily lead to a different ethics, 'political ethics'.

Another issue is party loyalty. It seems inevitable that democracy requires a party system. The politician is a member of an in-group, and is judged by his loyalty to that group. This is understandable, for he had the party's support when he was elected and perhaps hopes for similar support next time. But he may tell himself that that must take precedence over his personal beliefs, and this may apply even over such major issues as the decision to go to war.

Thus politicians have two major ways of justifying actions that would otherwise lead to a lack of congruency in the self-system. One is the belief in their own mission which can lead to excessive moral rectitude, and the other is seeing loyalty to party or country as overriding moral considerations.

Looking ahead, another problem, likely to become increasingly important in the future, deserves mention. Hitherto democracy has resulted in the reputation of political leaders depending to a large extent on how well they have looked after their country's interests. But the advances in science and technology in the last century make it essential for us to adopt foreign policies that are radically different from those widely accepted in the past. The wealth of the industrialized nations contrasts with extreme poverty in other parts of the world and the discrepancy is no longer acceptable. The world's resources are being consumed faster than they can be replaced. The devastation that can be caused by modern weapons makes it essential to avoid armed conflict. Policies aimed solely at doing the best you can for your own country, in essence selfish assertiveness at the group level, deepen and perpetuate the differences in the world, when the need and perhaps the actual trend is for all of us to be on the same side. A way must be found to align national interests with global interests, and national loyalty must be diluted by care for humankind. We have to learn to live together, otherwise we shall die together.

While we cannot predict what further advances science and technology will bring, we can be sure that they will result in changes in two directions: a higher quality of life for all with the enhancement of civilization; and the means to destroy that civilization and perhaps even the human race. Since the latter path is clearly unacceptable, we must learn to base policies on moral principles. If we are to have a future more free from tensions than has recently been the case, we must seek for more openness and honesty in both foreign and domestic politics, and a global outlook in foreign policy. It may well be that, as the importance of the sovereign state becomes diluted, domestic and foreign policies will merge.

I am aware that I shall be accused of idealism. But those who say that morality in politics is like pie in the sky should imagine what the world would be like if politicians simply disregarded all moral considerations. I have emphasized that the politician's path is sown with the seeds of conflict, with moral issues virtually always present and often pulling against traditional solutions, but I am not saying that all politicians are motivated solely by ambition, nor that all conflicts are insoluble. A first step must be to establish and abide strictly by the rules of law in international relations. Without this there will be unbridled national competition leading to anarchy in the world. And awareness of the sources of differences and the understanding of group biases provided by an evolutionary approach may help.

8

Ethics and Business

The commercial world poses ethical conflicts of great complexity at every level from the individual to the society. Many of the problems go unnoticed by the general public until there is news of gross mishandling or misappropriation of funds, as in the collapse of the Maxwell empire and the Enron scandal. As I write, the papers report that one of the giant oil companies, in trying to safeguard its markets, is pouring money into organizations that minimize the significance of climate change. It is partly to such events that the growing interest in business ethics is due: there are now innumerable books as well as a journal devoted specifically to it.

Major scandals such as these are usually due to the greed of a few individuals whose knowledge and position enable them to manipulate the system. Of more interest in the present context is the fact that the circumstances of business dealings lead many people to feel that it is justifiable to use standards in their business behaviour different from those to which they subscribe in their private lives.

The essence of the problem is this. We have seen that, in long-term interpersonal relationships, the satisfaction of both parties is a goal to be sought after in its own right: both parties must be satisfied if reciprocity is to continue (see Chapter 4). That is also the case with, for example, the village shopkeeper and a regular customer: the latter must be satisfied or she will not come back. But, to take a hackneyed and probably unfair example, the second-hand car dealer whose customer is unlikely ever to return may be tempted to bargain and to get as much as he can: the Golden Rule of Do-as-you-would-be-done-by is transformed into an ethic of legitimate competition, I-will-do-the-best-for-myself-because-I-know-you-are-trying-to-do-the-best-for-yourself.

Now consider a street full of second-hand car dealers in competition with each other: a customer goes from one to another seeing where he can get the best deal. Each seller must lower the price he asks or the customer will go elsewhere, and that is to the customer's advantage. Thus, while business dealings can involve behaviour contrary to the Golden Rule, business competition brings a more efficient market and better prices for consumers.

The problems are exacerbated further by the fact that business organizations are themselves complex, and are embedded in local, national, and often international environments: the people involved have conflicting interests, and every decision has diverse and often unforeseeable consequences. Loyalties conflict, and individuals must change their loyalties, for instance from the firm to their workmates to their families, as circumstances change.

Many economists would dispense with prosociality in the business world altogether. On this view, while prosociality is fine in personal life, in the marketplace it can destroy the benefits to consumers that the competitive marketplace brings, such as lower prices. From that it follows that we should all live in two worlds, the world of personal relationships and the world of the marketplace. If we were to use the standards we accept in the former in the marketplace, we would destroy it and lose the benefits of a competitive economy. As V. L. Smith puts it, 'the work that markets do and the unintended good we accomplish through them is completely foreign to our direct personal experience'.[1] For the competitive market to bring the benefits that Smith believes that it brings, we must transfer the trust that is necessary for personal exchange to the institutions that enable exchange and specialization to be extended to vast networks of strangers. This economist's view of the benefits of competition is thus concerned with a different level of analysis—with benefits to the population involved, not with the motivations of or consequences for the individuals concerned. We shall see later that competition can bring both benefits and harm, and also that fair dealing can bring its own rewards.

In discussions of business ethics 'Rational behaviour' is usually equated with self-interested behaviour. But we have seen that individuals do often help others even when it is against their own material interests. They do often share their resources with others when they are not compelled to do so. In the real world self-interested behaviour must be constrained if societies are to survive. Personal relationships between individuals and the welfare of society both depend on a proper balance between prosociality and selfish assertiveness.

This chapter is concerned with the constraints placed on the ethical behaviour of individuals by the demands of business. It uses skeleton scenarios lacking many of the complexities of the real world to emphasize the ethical problems that arise in business. This gives a rather negative picture, and many will rightly feel that it does scant justice to the many honest and conscientious businessmen, but my aim is to highlight the problems. In a final section I discuss some hopeful signs that ethics can play a more positive role in business.

DYADIC DEALS

Consider the simplest case, the selling of an object by one individual to another, and neglect the influences of and consequences on other parties. A first question concerns whether the object or service *ought* to be subject to monetary negotiations. There are some things for which it would be unacceptable to ask for payment in practically any culture—for instance hospitality to a friend in one's home. However, one would ask a price for one's house or one's car, even from a friend. In other cases payment is made, but may be considered wrong. Is it right, for instance, that money should be paid for sex, when some degree of coercion might be involved: the prostitute may be coerced by poverty?[2] Or that body organs be bought for transplantation, when the monetary payment can degrade or corrupt the good that might otherwise accompany the transaction? Or that places at a university should be allocated on the basis of likelihood that the recipient or his/her family would contribute to university funds?

Often the value of the object of the exchange cannot be objectively determined. If there is any point to the exchange, the buyer necessarily values the object more than does the seller. If the object is worth £10 to the seller and £20 to the buyer, they may agree on a price of £15. The buyer will then be paying less than the object is worth to him, and the seller will be getting more. Both would be satisfied and neither would be considered dishonest.

In the real world, each is likely to try to get the best deal he can for himself while satisfying the other, and each knows that the other is similarly motivated. In bargaining, the buyer will aim for less than £15, the seller for more. In these circumstances, where does honesty come in? If the seller aims for a price higher than the object's value to him, is that dishonest? And if the buyer goes away happy because he thinks he has got it for less than it is worth to him, is he being dishonest? In terms of the Golden Rule, neither is doing to the other as he would like to be done by. I have argued that the principle of Do-as-you-would-be-done-by is virtually ubiquitously applicable (though interpretations vary), and limits, or should limit, the nature of the precepts elaborated in a culture. The precept that everyone should do the best for himself would contravene this principle. And, in fact, few would accept a precept of licensed unlimited self-assertiveness in exchanges as a general principle.

But, in another way of looking at the matter, neither is being dishonest in any of these scenarios. Each, consciously or unconsciously, recognizes that, in the context of the business world, the seller is entitled to make a profit. The rules of the game are known to both parties: each expects the other to try to get the best possible deal for himself. While such a convention is recognized in the West, there is nevertheless a feeling that there is something slightly tainted about bargaining. In the Middle East bargaining may be almost a matter of honour. We saw in Chapter 1 that each culture evolves precepts to fine-tune the behaviour of its members: perhaps the precept of 'Every one should do the best for himself' can be seen as universally more or less accepted in the mini-culture of business ethics, justified at the personal level by the presumption that the other party is acting in the same way, and at the community level by the view that the market works better that way because consumers get a better deal.

This is, of course, the economist's view. Most dyadic transactions take place in a marketplace. Suppose that there are two or more potential buyers, each of whom values the object at £15 and will not pay more. Fifteen pounds is then seen as the market price. Such issues may have repercussions on price levels throughout the marketplace. Sellers cannot charge an excessive price, and may be forced to charge less than £15 by competition with other sellers, and consumers will pay a price lower than the seller might want to demand.

Thus even this simple case shows that the problems lie not just between prosociality and selfish assertiveness, but between the precepts or conventions of two different worlds—the everyday world and the business world. The participants can see themselves as acting according to the precepts that apply in the marketplace 'Do-the-best-for-yourself-because-he-is-doing-the-best-for-himself'. The outsider may see this as a distortion of the Golden Rule. And the business perspective has come to be accepted because 'the market works better that way', leading to cheaper prices for buyers, and just bad luck to those who lose out in the competition.

THE BUSINESS WORLD

Thus over many of the issues raised here, the competitiveness of the marketplace can provide excuses or reasons for abandoning everyday morality. For those who argue that the only responsibility of a business is to make profits, provided it stays within the law, competition in the marketplace determines what is fair, and the low prices that result benefit the consumer. Smith and many economists have argued that behaviour that we might see as immoral at the individual level can benefit the collective, so the morality of actions at the individual level should be assessed also by their consequences at the group level. But it is necessary to pursue some of the complications in the real world.

Most transactions take place in a marketplace with competing buyers and sellers, and the seller employs a number of salesmen (or, in the case of a manufacturing business, employees of varied sorts). Many questions arise. Employers strive to maximize efficiency and profitability, employees seek to maximize their wages, working conditions, and security. Is the salesperson paid a fair wage? What is fair? Does the employer provide adequate safety precautions at work? What are adequate? Is the employer prejudiced ethnically or religiously in choosing employees? What is morally right or wrong may differ from what business ethics permits, let alone what is legally right or wrong. For instance, the limits of what are adequate safety precautions may be equated with the legal requirements, rather than with what is morally proper.

We must also ask what instructions are given to salesmen as to how far they should go to make sales. In particular, should they lie? At what point is exaggerating the merits of the item for sale to be considered as lying? If the salesperson exaggerates, is it the management or the salesperson or the ethic under which he works that can be seen as responsible for the dishonesty?

There are many other issues, such as the source and nature of the article being offered for sale. Did it 'fall off the back of a lorry' or was it acquired honestly? We could also ask whether those who made it were paid a fair wage and whether its manufacture involved resources extracted with damage to the environment or involved unnecessary and cruel experiments on animals. Yet another issue is whether it was advertised fairly. Such questions are almost endless.

There may be both competition and cooperation between the employees. Each employee will be affecting both his peers and the management if he calls in sick without justification, or cheats on his expenses, and the employer will be treating the majority of employees unfairly if he overlooks the misdeeds of one of them to avoid a disruption to the business. In other words, loyalty to the collective is involved as well as dealings between individuals. In a well-run firm, feelings of loyalty may be shared by all the employees; but in a poorly run one there may be an unspoken agreement among the employees to see what they can get away with. Loyalty and cooperativeness will depend crucially on the employees' view of the firm (see below).

The distribution of the resources available for paying the employees can be a major issue. The employees should feel not only that they are being paid adequately, but also that they are being paid fairly. As G. C. Homans, properly regarded as the founder of Exchange Theory[3] (see Chapter 4), put it, 'For with men the heart of these situations is a comparison.'

Consider the criteria of fairness discussed in Chapter 4: each is capable of justification in moral terms, but pragmatic considerations enter and must be considered in each individual case. A simplistic view might be that all employees should be paid equally.[4] Ideally equal pay could result in the minimization of tension among the employees, because individuals tend to compare their pay with that of others. However, that runs into problems. An employee may feel 'Why is he paid as much as me when he works less hard, is less skilled, or has such an easy and pleasant job?' And, human nature being what it is, each employee may do the least amount of work that is compatible with his keeping his job. The efficiency of the business would suffer, it might fail in competition with other enterprises, and the interests of neither employer nor employees, nor indeed the general public, would be served. Considerations at the firm and societal levels make the simple solution of equal pay impracticable. It must also be remembered that, if it is a matter of pay increases where there are already differentials, an equal increase to all workers would be less meaningful to the higher paid workers. Should all be given the same percentage increase?

A second possibility is to apply so-called Marxist justice, paying employees according to their needs. On this view, an employee with many children to support might be paid more than one with none, and an unmarried individual more than one with a wage-earning spouse. Though this has been tried out in Japan, there can be no end to the difficulties of assessing need, there are likely to be large differences in the assessments made by employer and employee, and jealousies between employees are likely.

A third possibility is that employees' pay should be affected by their 'investments', in the sense of what they are invested with. In the traditional army, for instance, it was not questioned that officers, recruited from the upper classes, should be paid more than lower ranks. Again, for centuries it was taken for granted that women should be paid less than men because of an archaic prejudice that men have higher value: this entrenched set of prejudices was challenged only in the mid-twentieth century and still lurks in some places. It is certainly unfair if women lose out over pay or promotion because of the possibility that they are or will become pregnant. Fortunately such prejudices are disappearing from society, but too slowly.

However, skills and training count as investments. Skilled workers are paid more than unskilled, and those with previous training more than those without. This, of course, is the result of market forces: skilled and trained workers are less abundant than unskilled and untrained ones, and contribute more to the profitability of the business. On this view, pay should be determined by the worker's contribution to business efficiency and by the availability of skilled workers: business efficiency may then ensure job security for all the employees as well as being in the public interest. However, pay in accordance with skills carries with it the consequence that unskilled jobs that require heavy labour, often in unpleasant conditions, are less well remunerated than office jobs involving sitting in an air-conditioned office facing a computer screen: dissatisfaction in the workforce and a resulting loss of efficiency may result. This is especially the case if the bosses, who may be perceived as having ideal working conditions and doing little as they sit behind their desks, award themselves large salaries on the grounds of their responsibilities.

That brings us to the possibility of compromise, namely recognition that the pay of employees should vary both with the nature of their work and their productivity. This would be justified because, on the one hand, some sorts of work are more attractive than others and, on the other, productivity is likely to be related to the skill and industry of the employee. Pay according to productivity provides an incentive to work hard, and can therefore be justified on pragmatic grounds. However, payment according to productivity can be unethical if it leads to the employee working excessive hours.

Lower paid workers may accept that those with more skill or responsibility than themselves should be paid more. In exchange theory terms, each worker compares the ratio of his rewards to his costs with the ratios of his fellows. What matters is that pay differentials should be *seen to be* fair, and, if skill is in short supply, it may be accepted that pay differentials are determined by the market forces operating.

In every one of these cases, ethical and pragmatic concerns are intertwined. The ethical arguments in favour of equality fail to meet the demands of the real world for efficiency. Efficiency is forced on the firm by the demands of competition and, from the point of view of the general public, it is desirable that the firm should be efficient in delivering the goods that the public requires and that the cost should be as low as possible. Payment according to need is impracticable, and would in any case lead to the same problems of inefficiency. Payment according to investment is indisputably unfair, except in those cases (skill and training) where it leads to greater efficiency: the differentials are then determined by market forces.

There is, however, another matter. Because firms do not operate in isolation, but in competition with other firms, employees may have some choice as to which firm they work for, and pay is not the only issue. Employees seek good working conditions, fair treatment, safety, security, and so on. Employees may also prefer to work for firms with a high moral reputation. There is more to job satisfaction than the size of the pay packet. Evidence from wage differentials has been used to assess the attractiveness differentials of jobs with differing ethical contents.[5] For instance, Cornell graduates entering socially responsible jobs accepted lower salaries than those taking employment in for-profit firms, academic ability and other variables being controlled. Again, public interest lawyers tend to earn less than corporate lawyers, and expert witnesses testifying that smoking could have deleterious effects on health were paid less (or nothing at all) than those testifying for the tobacco industry. Many employees seek also for a situation where they will not have to live their lives on ethical principles that differ between work and home.[6] A firm that looks after its employees will be preferred to one that pays them the minimum it thinks it can get away with. Employees see the firm as their in-group, good treatment of employees enhances their loyalty to the firm, and this augments cooperativeness and efficiency.[7] Thus it may be in the firm's interests to pay its employees well, to provide them with amenities, so far as possible to give them scope for some autonomy, and generally to look after their welfare as an incentive to working well: prosocial behaviour by the firm has a pragmatic basis in self-interest. However, the emphasis placed on employee loyalty is changing in some industries. In information technology and other enterprises that depend on innovation, employees are not expected to stay long with the firm. Indeed, it may aid innovation to have a frequent input of new blood in the team.

Perhaps employee loyalty is best assured if the firm is in effect owned by its employees, the executives being their elected representatives. The John Lewis Partnership in the UK is such a firm, and Scott Bader is another. The latter is discussed later in this chapter.

EXECUTIVES AND SHAREHOLDERS

So far we have assumed that the owner of a business was a single entity. Most businesses are small, so that is usually a justifiable assumption. But it is the large businesses, employing hundreds or thousands of individual workers and having a hierarchical command structure, that get most attention and where the ethical issues are most important. Simplifying greatly, the management of public companies is usually in the hands of executives, who are responsible to the shareholders. The shareholders may be largely institutions. (We leave aside the extent to which there are specialist managers in charge of public relations, advertising, sales, and so on.) In a large business the powers of the executives are controlled by a board of directors. Of course, all the ethical problems discussed so far still apply, but now additional ones must be considered.

Such a 'corporation' may be seen as an entity, owned by the shareholders, or as an individual, with the right to own assets, and responsibilities to outsiders. On the former view the company has no conscience, and companies will act morally only so far as their actions are constrained by law. Some governmental interference is necessary because unfettered competition inevitably infringes human rights and acts against a fair distribution of resources.[8] But because many laws depend on establishing a particular moral culpability, constraints on company activity are sometimes difficult to enforce. And the question arises, how far does government legislation unfairly constrain individual company rights and limit the supposed benefits of a free market? Thus restrictions on cigarette advertising may be in the public interest but against the interests of manufacturers and shareholders, and even of the smoking public.

It is, of course, important that the lawmaking body should be independent from the corporations. Governmental goals may run contrary to business interests: the government is (or should be) concerned with the public good, while a business is concerned with its own economic interests. Of course, governments still encourage and give special support to businesses in which they have an interest, such as armament firms, and that may or may not be in the public interest.

On the second view, the company is an individual and can be held morally responsible for its actions, even if they are within the law, if harm is caused to others. In practice it is the executives who make decisions on behalf of the company who hold the responsibility, though it may be very difficult to pin down just where responsibility lies.

The responsibilities of the executives are multiple and involve all who have a stake in the company. Simplifying greatly, stake-holders include the shareholders, who rely on the company to maximize profits, the employees who must have fair pay and conditions, and the general public, who desire lower prices in the marketplace thereby implying lower profits for the firm, but also goods of reasonable quality that are safe to use. In addition the executives may have responsibilities to smaller firms who supply them with raw materials or components. The executive's problem is, in essence, who to count as members of the in-group. Their position is thus far from simple. In striving to increase business profitability they may see themselves as acting properly in terms of business ethics, though their actions are contrary to everyday ethics.

As we have seen, attention to the employees' pay and working conditions may have positive consequences for the company's efficiency and profitability, and thus be compatible with the executive's duty to the shareholders.[9] On the other hand, too much attention to the demands of the employees could reduce the proportion of the company's profits due to the shareholder-owners. The executives are the employees of the shareholders and they may be held responsible for their past year's actions at a share-holders' meeting. They are therefore likely to be caught between conflicting loyalties.

Where executives determine their own salaries, do they really deserve the large salaries they are paid, and are such salaries necessary to attract the best people? In principle their salaries should be determined by the shareholder-owners of the company, who may operate through a board of directors. Directors may be appointed by the employees, whose interests they are supposed to represent on the board. The board then can, in principle, control the executives' perks, and monitor and constrain their actions through the year. In practice this does not always work well, since those appointing directors may in practice have little knowledge of the requirements of the job or the merits of the candidates, directors may have little direct involvement with the company, and in some cases executives may choose the directors or be themselves directors. A related issue concerns the use of bribes. Recent studies have shown that this is related to state poverty: the giving of bribes is much more prominent in poor countries than in rich ones.[10]

The corporation, and thus the executives, may be held respon-sible for any damage to the public or to the environment that the corporation's products, or the processes involved in their man-ufacture, may give rise to. Thus car manufacturers must ensure that their cars are safe; pharmaceutical companies must ensure that the medicines they produce are effective for the specific purpose for which they are intended and are without harmful side-effects. Too often, however, businesses are able to find their way round the law. For instance in the USA the Manville Cor-poration produced asbestos which was responsible for death or injury to thousands who were in a position to sue the company for damages: the company filed for bankruptcy as a means of avoiding having to pay compensation. However, the firm later reorganized and set up a trust fund to compensate the victims, the shareholders bearing most of the cost. In an even more widely publicized case, a leak at Union Carbide's factory in Bhopal, India, led to an estimated eight thousand deaths immediately and tens of thousands since: virtually no compensation has yet been paid. In addition, the environment, and the rights of future generations, have no voice, and are easily disregarded by the executives: their case must be argued by members of the public.

While no doubt most executives do their best to reach optimal solutions to the conflicting demands placed upon them, corporate profitability often runs contrary to everyday ethical considerations. For instance the scale of deaths resulting from employment is often not appreciated. The number of homicides in the UK recorded each year by the police was only about a quarter of the deaths due to industrial accidents and diseases. Some of the latter can legitimately be attributed to 'accident' or managerial ignorance of the potential dangers inherent in the process or product in question. But most accidents are avoidable if proper precautions are in place, and potential dangers can be foreseen if enough trouble is taken.

In the UK the extent to which company directors are legally responsible for injuries to their employees or the public is still confused, though a parliamentary decision is pending. In the last eight years nearly five thousand people have been killed in 'workplace incidents'. There have been spectacular disasters like the Piper Alpha explosion, and the Southall rail crash from which company directors have been able to walk away without penalties. While a change in the law has been under discussion for many years, a firm cannot be prosecuted for manslaughter unless it can be proven that one of the directors was personally responsible. Interestingly, a director can be disqualified and imprisoned for gross breach of duty towards the shareholders' investments.[11]

Corporations may also be responsible for depriving, usually indirectly, citizens of resources. For instance, multinational drug companies may charge the National Health Service more than alternative sources would charge for effectively the same product. Corporations may use unfair means to defraud competitors (e.g. by espionage, bribery, price-fixing, etc); governments (by tax evasion, or by using corporate funds to gain political favours, as was revealed in the Watergate scandal); employees (by underpaying, inadequate conditions of employment); consumers (by fraudulent advertising, adulterating their products); and the public (e.g. by pollution). Box[12] has listed many other examples, which he describes as corporate theft. Inevitably but regrettably, many of the decisions that executives make are probabilistic, for instance balancing a probable increase in sales against a probable increase in number of cases of lung cancer, or in number of complaints that an advertisement elicits with its effectiveness in increasing sales.

Corporate crime is, of course, immediately due to the actions of executives and others in similar positions. The executives may be doing what they see as right for the firm, seeing corporate profitability as their duty but sailing too close to the wind, underestimating the dangers in the industrial processes for which they are responsible, or overestimating the reliability of the products. Less often, they may see profitability as legitimately overriding the demands of ordinary morality. Lower ranks in the company may conform because their employment and/or preferment depends on their furthering the company's interests, and later assimilate the organizational morality as part of their own world view.

It can also be argued that firms have duties to their competitors. While cooperation between firms is forbidden in order to preserve the benefits of the marketplace for the consumer, some forms of competition, such as bribes, price-cutting, misrepresentation, hacking into the rivals' communications, and so on are rightly considered as improper and as unfair competition. While fair competition is seen as being in the public interest in keeping down costs for the consumer, in fact unfair competition could be equally so. But competition, fair or not, can lead to increasing wealth differentials, and it implies a devil take the hindmost attitude. The result of competition may be the elimination of a rival: this may or may not be good for consumers, but is unlikely to benefit the employees of the less successful firm. Putting them out of work may or may not outweigh the general benefits of fair competition.

Of course ethical entrepreneurs who try in various degrees to operate on everyday ethical principles do exist, but a study in the Far East of entrepreneurs who tried to operate on Confucian principles, and to resist the demands of instrumental rationality, found that their efforts were costly.[13]

Many investors feel that they should put their money only into firms that do not infringe their moral principles. Issues considered here include arms and tobacco manufacture, pollution of the environment, unfair treatment of employees, and animal testing of the products and, on the positive side, use of renewable sources of energy, donations to charities and to political parties, and effects on local communities in Third World countries. Sadly, the data on the financial return on ethical investments are decidedly mixed. Furthermore, the demands of ethical investors have a complex relation to corporate behaviour.[14]

In theory shareholders can influence a company's policies through their votes at shareholders' meetings, though this is always difficult because ethical investors are usually in a minority. However, they had a success when Barclays Bank closed its operations in South Africa during the Apartheid era, apparently as a result of the adverse publicity consequent upon the protests of shareholders and customers. In practice it is not easy to discover whether a firm's enterprises are entirely ethical, as most corporations have diverse interests.

A PUBLIC ROLE

In the preceding paragraphs, the interests of the general public have entered at several points. In an industrial society, the customers depend on enterprises of one sort or another for their needs—food, medical care, transport, entertainment, and practically everything else that makes for civilized living. Thus the public has an interest in whether or not firms fulfil their functions and in whether they do so efficiently. In a sense, it is the sum of competing firms in which the public has an interest, for if one goes under and the rest still produce the goods, the public may not suffer.

On these grounds, therefore, a firm or industry has a moral obligation to the public, as well as to its employees, its suppliers, and so on. But the possible complications are almost infinite. A firm's moral duty to the public may run counter to its business interests. For instance, a difficult situation arises when certain parts of a firm's operations become unprofitable, such as when country bus services or post offices are so little used that the costs of running them are not covered. Should they be kept open, perhaps under public ownership where profitability is less of an issue, in the interests of the village dwellers, or closed in the interests of the profitability of the enterprise as a whole? Furthermore, how much the firm charges its customers may affect the wages of the employees. Thus owners, employees, and customers have conflicting requirements, and in some cases, at least, the issue is both a moral and a pragmatic one.

Customers may weigh low cost against ethical standards: during the Apartheid era many Europeans refused to buy South African produce. Customers may also prefer to patronize a firm known to treat its employees well, or one that does not depend on cheap child labour in another country. The ethical demands of the public can affect what the firm does. Pharmaceutical firms have been especially affected, and some have taken steps to improve their image. Thus the developing world does not offer a commercially profitable market for many of the medicines that its citizens need, but some international enterprises, such as GlaxoSmith Kline and Pfizer,[15] claim to be devoting considerable resources to research on medicines and vaccines for malaria, HIV/AIDS, and other diseases, and to be making the drugs available at prices affordable in the countries in which they are needed. Similarly, such firms claim to use the minimum number of animals in research, to adopt environment-friendly policies, and so on. One could of course argue that such moves are ultimately self-serving and thus self-assertive rather than prosocial, but does that matter? They indicate that the limitations of 'pure' capitalism are being perceived by big business.

While in general the limitations imposed by the law are less extensive than those that are morally acceptable, there are times when the relations between legal and ethical requirements are complex. For instance, American Cyanamid prohibited pregnant women from working in toxic areas to protect their foetuses, but were later prosecuted as the prohibition was seen as a form of sex discrimination.[16]

In summary, the role of competition is more complex than appears at first sight. To an economist, a competitive market is one with no barriers to entry for new firms and in which no firm or group of firms has the ability to manipulate the market price. The ideal competitive market of the textbook is one in which there are many producers, each using best-practice techniques and each making no more than 'normal' profit. If one firm has market power, it could drive prices too high and output too low in order to make excessive profits: innovation is also likely to be slower. It is this that 'competition law' strives to prevent. To an economist a competitive market is what is wanted: the extent to which it is the best way to organize social resources depends on a number of factors, including the nature of the goods concerned. But the success of one firm may spell failure for others and devastation for its employees. It is, for instance, difficult for 'corner shops' to compete with supermarkets, and the juggernauts of the retail business are left to compete with each other. More generally, economic competition leads, apparently inevitably, to greater discrepancies between rich and poor. In recent years the salaries of chief executive officers have gone from forty times to four hundred times the average worker's pay. This is not something to be accepted as inevitable: wealth discrepancies are one of the strongest correlates of violence in and between countries.[17]

ENVIRONMENTAL ISSUES

Two issues arise from the impact that commercial activities have on the environment. First, there is the damage to the environment. This can take many forms, from pollution of the atmosphere to the felling of forests and disposal of mining wastes. The effect may be distant in time and space from the activities that caused it. A second and more subtle problem arises from depletion of the earth's resources.[18] Dasgupta and Mäler have pointed out that most indicators of social well-being (e.g. GNP, life expectancy at birth) do not take into account the impoverishment of natural resources that may result from commercial production processes. Resource degradation may be occurring, but there are seldom any indicators of whether it is approaching a state where current activities will become unsustainable. It is easy to forget that the world's resources are limited.

Excessive exploitation may have immediate consequences on the environment, and if the firm is forced to move elsewhere the local population will suddenly lose an important source of income. An example is provided by Shell's attempts to extract oil in the Niger delta. One prominent activist who had taken a stand against the company was executed by the Nigerian Government in 1995. It is reported that the Ogoni people 'drove Shell out of their part of the Niger delta' because they were getting no benefits and much pollution from the company's activities. In some cases spilled oil caught fire, destroying mangrove forests and farmland. It also polluted fishponds and drinking water. Apparently, Shell has not been back to the community, though the installations are still there.[19]

OPERATING ACROSS CULTURAL BOUNDARIES

Further difficult questions arise from the fact that many companies now operate in several cultures. Satellite operations are established, often in the Third World, because labour is cheaper there or raw materials are locally available. But should a Western firm pay the very low wages that are acceptable in countries where labour is cheap? Is it proper to profit from the lack of governmental restrictions on the length of the working day in other countries? How should employees of Western firms treat women in traditionally Islamic countries? And how should executives interact with local governmental representatives and officials? Regrettably, such decisions are often made on purely pragmatic grounds. This has resulted in disruption of local customs and infringements of human rights.

But ethically the problem may be even more complicated. For instance, if the local government supports laws that are not compatible with the Golden Rule, perhaps permitting racial or class discrimination, should the foreign firm be there at all? Attempts have been made to formulate codes of practice acceptable in different cultures, including codes based on the Golden Rule.[20] As noted already, there is sometimes a case for rephrasing the Golden Rule as 'Do-to-others-as-you-believe-they-would-wish-to-be-done-by', for there is still just a possibility that some Islamic women, brought up in their own traditions, might prefer to be treated in the way to which they are accustomed, and there is no knowing the extent of the repercussions of any attempt brashly to impose Western values. All Western attitudes and values are not necessarily best for everyone, and sensitivity must be a key word for companies operating in other cultures.[21]

As emphasized earlier in this chapter, we are concerned with the conflicts that arise and with where things may go wrong. Many will see this as resulting in too black a picture. While the activities of any company are likely to have both positive and negative consequences, there are undoubtedly some companies, like Unilever, whose record is said to be predominantly positive.

WHISTLE-BLOWING

When an employee feels that his firm's activities are immoral or illegal, should he publicize his misgivings? To do so would be a breach of loyalty to his employers, and might mean breaking an undertaking. It will certainly mean the loss of his job, with consequences for his family, and may lead to prosecution. The alternative is the guilt of contributing to an improper enterprise. A frightening example was recently reported in the media. Apparently, a few years ago Franklin, a doctor, was hired by a pharmaceuticals firm in a 'medical liaison' position. His job was to promote a drug that had been approved for the treatment of epileptic seizures. He found that in practice he was expected to recommend it also for a range of other complaints. When he found side-effects in some of the children treated he was ordered not to tell doctors. He also found that doctors were paid cash and elaborately entertained so that new uses for the drug could be suggested to them. The firm, which had been taken over by Pfizer, was eventually sued successfully, and fined a $240 million criminal fine and $152 million to state and federal healthcare authorities.

Franklin suffered through seven years of legal wrangling, with his career in tatters, and became a recluse. In the end he received $26.6 million for his role, a large reward. But one must set that against what he suffered and the devastating consequences he would have suffered if the case had come out the other way. The case provided a spur for other successful prosecutions of pharmaceutical firms in the USA.[22] In many other cases where whistleblowers have been subjected to physical attack, culpability has not been established.

SOME INTERIM CONCLUSIONS

It is easy to caricature the business world, and this chapter has given only a glimpse of the ethical problems involved. But this glimpse strongly suggests that traditional ethical values are distorted, even in simple person-to-person bargaining, by the situation. It is as though the usual balance between selfish assertiveness and prosociality has been shifted in favour of the former, with the prosocial reciprocity of person-to-person interaction in everyday life superseded by a business ethic of mutual understanding that each must do the best for himself. This is in accordance with economic theories that assume that individuals will always seek maximum gain, with 'rational self-interest' as the guiding principle. Economists support the competitiveness of the market, maintaining that it leads to lower prices and greater efficiency. That greed and the forces of competition can be excessive is apparent in the way that global warming proceeds, rivers are polluted, and forests are felled. In addition competition leads to greater differences between rich and poor, and that in turn may lead to societal violence and other undesirable consequences.

However, prosociality is always potentially present. Even the most hard-bitten salesman may be haunted by guilt, perhaps rationalizing it by referring to the customers as gullible suckers. Pressures for fair dealing are made evident in the demands of employees, customers, and, often, the government. These pressures are leading some corporations to behave 'altruistically', though the decision to do so is based on pragmatic grounds, for the firm may profit from ethical behaviour. It may also lead to the firm trying to have both worlds, falsely presenting itself as upholding moral principles.

It must be acknowledged that it seems almost impossible for an organization to behave ethically in every respect: even the Norwegian firm Norsk Hydro, a company that has set an outstanding example in taking its ethical responsibilities seriously, has been criticized over some aspects of its operations.[23] In the complexity of most business situations many decisions inevitably involve conflicting loyalties.

These matters have become more important because many people believe that, with the demise of communism as a potential world force, capitalism and its associated emphasis on competition no longer have a realistic alternative. If that is so, and I am not implying that communism was any better, ethical dealing must be sought amid the constraints of the capitalist system.

IS BUSINESS NECESSARILY RED IN TOOTH AND CLAW?
CORPORATE SOCIAL RESPONSIBILITY

Free-market capitalism can benefit the general public, but it can also lead to social injustice, inequity, and increased differences between rich and poor. In addition, any commercial product may have adverse consequences for society, and if an entrepreneur manufactures a product that has adverse consequences on society, it may be difficult for him to stop. If you make weapons, it is necessary to make more and better ones if you want to stay in business.[24]

There is, however, hope for the future. Not all firms seek to maximize profit and neglect their social responsibilities. Many groups of businesses elaborate their own codes: for instance the Association of British Travel Agents has a code of practice for travel agents. That such codes are the result of enlightened self-interest is another issue. And many large companies are beginning to review their responsibilities, questioning the supremacy of the profit motive. Corporate Social Responsibility is coming to receive the attention it deserves as an essential element in organizational structure and planning. Corporate responsibility is being taken beyond the effects the organization may have on the environment to ask whether an industrial project can both advance human rights and lead to sustainable development.[25]

Welcome developments come from the attempts of some non-governmental organizations (NGOs) to reform market systems rather than simply confronting them. For instance, Amnesty International has helped train executives of companies working in conflict zones in human rights issues; and Greenpeace has worked with firms to promote a renewable energy product in the UK.[26]

Recognizing that business can be an enormous power for good or evil, at Davos in 1999 the UN Secretary-General pointed out that the rate of organizational change in the world exceeds the ability of social and political structures to adjust. There is an urgent need for cooperation between the business world, its stakeholders, national and supranational authorities, and NGOs. This involves recognizing not only that the use of non-renewable resources must be taken into account in calculations of sustainability,[27] but also care for issues of culture, responsibility, workers' rights and individual welfare, and for the long-term effects on the local communities affected by companies' operations. The emphasis is on the long term not only because of the depletion of resources, but also because development may involve disruption of a society's social structure and ethical system. It may be necessary to help local communities through the adjustments necessitated by the intrusion of the new enterprise, and to do so with an eye on the communities' long-term good.

Many corporations are seeing their responsibilities as extending beyond their own employees to those of their suppliers. This may mean no exploitation of child labour, no sweatshops, and respect for human rights. Companies can also promote human rights on the spot—by making clear their attitudes to abuses, and practically by helping to support hospitals and schools. The Norwegian Government, in collaboration with Amnesty, has shown how this can be done in China, Indonesia, and elsewhere.

It is becoming recognized that companies operating in a Third World country must try to benefit the community by creating wealth, enhancing local well-being, and avoiding irreversible harm to the environment. It is not enough just to promote temporary economic development in an 'under-developed' country. They must also ensure that that wealth is used properly and that they are not party to corruption, political repression, or abuses of human rights. However, they have to do this while minimizing the cost to the stakeholders on whom they depend—the employees, the community, and the shareholders.

Perhaps this is best done where the employees are the shareholders. Scott Bader, a chemical firm, was formerly Quaker-run. The shares are held in trust on behalf of the employees. Employees are expected to adhere to a common set of principles and maintain a reputation for honesty, integrity, and quality in all dealings. The firm does not deal in materials that might be used to make weapons. The company has grown from small beginnings into a well-established international company. Because the shares cannot be sold, they have to rely on their profits for reinvestment and expansion, but they have succeeded in establishing branches in South Africa, France, and Croatia. It has made considerable contributions to local charities.[28]

In the short term, it need not be a matter for concern that companies' motivations for taking Corporate Social Responsibility seriously are partly self-interested. In the end, the recognition of Corporate Social Responsibility must depend on a change in attitude and on education of new entrants into the business world. Business schools must no longer teach that the aim of businesses is solely to maximize profit to shareholders. Some firms do in fact include a programme of education in corporate social responsibility for their employees.[29]

CONCLUSION

From two-person bargains in the marketplace to the dealings of multinational corporations, profit dominates the business world. Bargaining implies a precept of Do-the-best-for-yourself-because-he-is-trying-to-do-the-best-for-himself. This may be diluted to some extent by the need to please the buyer or to maintain a good relationship with the other party or a reputation for fair dealing amongst third parties, but the precept has become accepted as properly guiding business dealings in many contexts. This in no way implies that there are not many honest businessmen and women who do their best to cope with the constraints imposed by the business world.

In businesses from small firms to large corporations, competition between producers or suppliers exacerbates the situation: each entity must strive to maximize its profit. The complexity of the business world places virtually impossible ethical demands on those who make the decisions. Their task is made more difficult because they must always see their own firm as the in-group in competition with others, and they are confronted by incompatible demands from groups inside and outside the firm.

Economists see competition as beneficial to the consumer in making businesses more efficient, bringing lower prices, and so on. This makes it easier for the individual businessman to see his behaviour as governed by an ethic that justifies selfish assertiveness not only on the grounds that the same precept applies to all those involved, but also that the public is better served. But competition involves losers as well as winners, and increases the differences between the haves and have-nots. Many outside the business world feel slightly uncomfortable with business ethics, but accept them as part of the world we live in. But, as globalization brings changes to the business world, it is proper to be vigilant, and to ask whether it is possible to retain the good and minimize the harmful consequences of competition. Firms that embrace and practise a goal of Corporate Social Responsibility are showing the way.

9

Ethics and War

The preamble to the United Nations Charter states:

We the peoples of the United Nations determined to save succeeding generations from the scourge of war, which twice in our lifetime has brought untold sorrow to mankind

There can be no question that, if we remember our humanity, war is wrong in principle and nearly always wrong in practice. It involves inflicting, or trying to inflict, harm to others that one would not like to have done to oneself. While war may strengthen bonds within the in-group, it contravenes the principle of prosocial reciprocity so far as the out-group is concerned. But we have seen (Chapter 2) that our morality emerged in situations in which successful competition with other groups was to the individual's advantage. For several thousand years, at least, war has therefore been accepted as a natural way to settle disputes between groups. Fortunately, the world is now beginning to see that what is natural is neither necessarily morally right nor pragmatically wise. Even without modern weapons, the horrors of Rwanda show that war is neither a sensible nor a practical way to settle disputes. Given modern weapons, President Kennedy put the matter forcefully: 'Mankind must put an end to war—or war will put an end to mankind'.[1]

This chapter attempts to specify the important factors that have led, and are leading, to changes in attitudes to war and with the manner in which war requires both national governments and individual combatants to bend the rules. It is concerned with organized war within or between states, rather than violent encounters between small groups, as found in chimpanzees, hunter-gatherers and slash-and-burn farmers, and modern street gangs. These are mostly opportunistic clashes in which males cooperate to kill or injure rivals perceived as weaker than themselves.[2] Although similar incidents occur in modern warfare, war involves hierarchically organized groups in which much of the violence results from obeying orders, or from duty to superiors, comrades, or ideals.[3] Duty in wartime may be contrary to precepts accepted in everyday life.

In the early sections of this chapter war is examined from the point of view of the politicians and others who initiate war. A later section discusses the conduct of war on the ground, emphasizing the moral conflicts to which combatants are subject. In the concluding section some steps towards the elimination of war are suggested.

A BRIEF HISTORICAL PERSPECTIVE

Perhaps because they recognized fully the horror of war, St Augustine in the fourth century and St Thomas Aquinas in the thirteenth century expounded variants of a 'Just War' doctrine, limiting the right to make war to purposes that were considered 'just'—a matter about which there could be considerable dispute. In so far as the 'just war' idea implies that any war could be just, it is a dangerous doctrine. In any case, with the rise of nation states it was neglected, and the right to go to war was regarded as unlimited.

Some ineffective attempts to limit the right to go to war on basically moral grounds were made before World War I, for instance in the second Hague Peace Conference of 1907. The turning point came when, after the terrible slaughter in that war, the League of Nations attempted to restrict the right to go to war, thereby making it unequivocally a *legal* matter—though, of course, many continued to consider it as also a moral matter. Henceforward the right to go to war could be seen as governed by law, and law became equated with morality. Various attempts were made to strengthen its provisions, including the Kellogg-Briand Pact in which most of the major states renounced war as an instrument of international policy. But the League lacked the ability to enforce its prohibitions, and was seen to fail dismally when Italy invaded Abyssinia (Ethiopia) in 1935.

The UK's declaration of war on Germany in 1939 on the grounds of its invasion of Poland was seen as justified by the obligations of a recently made treaty with Poland, and on the grounds of self-defence, the German invasion of Poland being an extension of a series of territorial incursions which could be seen as likely to continue indefinitely. (In retrospect, after the revelations of the Holocaust, the declaration of war on Nazi Germany could be seen as morally justified by combating a greater evil, but in 1939 the extent of the Nazi ill-treatment and murder of Jews and Gypsies was not widely appreciated.) After World War II it was accepted at the Nuremberg trials that the principles in the Kellogg-Briand Pact were part of customary international law, and the Nazi leaders were convicted on the grounds that they had conspired to wage aggressive war.[4] In 1946 the Charter of the United Nations outlawed resort to aggressive force unless authorized by the Security Council.

Unfortunately, differences between countries and between groups within countries do occur, and those involved still fail to recognize that war, with certain rare possible exceptions, is not only morally and legally wrong, but also pragmatically usually a bad choice. Violence breeds violence, and the consequences of war are unpredictable. In the following sections we consider first the morality and legality of going to war (*jus ad bellum*) and, second, the question of morality in war (*jus in bello*).

DECLARING WAR

Differences of interest, both between countries and between groups within countries, are certainly bound to occur so long as the world's goods are shared inequitably, and even if they were shared equitably, ideological and other issues could arise. With the growth of international law, the immorality of resolving such disputes by military means has come to be equated with illegality. Article 33:1 of the UN Charter enjoins the parties to such a dispute to 'first of all, seek a solution by negotiation, enquiry, mediation, conciliation, arbitration, judicial settlement, resort to regional agencies or arrangements, or other peaceful means of their own choice'.

Disputes do not necessarily escalate into violent conflict, though with the human propensity for self-assertiveness such an outcome is always a possibility. The Security Council of the UN, under Article 34 of its Charter, is entitled to investigate any situation that might endanger international peace and security, and recommend appropriate procedures including armed force (Article 42). Except in self-defence, states are prohibited from using force or threat of force against 'the territorial integrity or political independence of any state or in any other manner inconsistent with the Purposes of the United Nations' (Article 2:4). Only the Security Council has the right to use or authorize force.

Times have changed since 1946, and the unanimity between the major powers that was expected then has never materialized.

Effective UN action to prevent war has been thwarted on many occasions because a member of the Security Council used, or threatened to use, its veto to prevent it. In addition, lack of adequate support from member nations has meant that the UN has not had the ability to intervene by force when force was needed. In Korea, and before the 1991 Gulf War, it had to entrust its mission to member states. In Rwanda, the UN Force was simply inadequate, and was withdrawn. Similarly the protection of a number of cities in the former Yugoslavia collapsed in part because forces adequate for the task were not available.

There have also been three major legal problems with the wording of the Charter.[5] One concerns the interpretation of Article 2:4, which prohibits the 'use or threat of force against the territorial integrity or political independence of any state'. For example, in 1949 the UK, wishing to remove mines, similar to those that had damaged its ships, from Albanian territorial waters, argued that its action was not intended to absorb Albanian territory or to subjugate Albania, and was therefore not an infringement of Article 2:4. However, the UK's conduct was condemned as a form of forcible self-help inconsistent with the purposes of the UN. In invading Grenada in 1983 the USA claimed that it had acted on the invitation of the legitimate government and that it was acting to protect US nationals. In fact the invitation came only from the Governor General, a post that carried no executive powers. The invasion, which resulted in the overthrow of the old government and the installation of one more congenial to the USA, was strongly condemned by the General Assembly. Again, the USA justified its armed intervention in Panama in 1989 as involving self-defence of its own nationals and defence of the Panama Canal under the 1977 Canal Treaty. The intervention was condemned by the General Assembly. However, other cases of states forcibly intervening to protect or rescue its nationals have not raised objections in the UN.

A second and major legal problem has arisen from Article 51, which provides that there is 'an inherent right of individual or collective self-defence if an armed attack occurs against a Member of the United Nations, until the Security Council has taken measures necessary to maintain international peace'. The words 'armed attack occurs' in Article 51 seem fairly unambiguous, and would seem to cover the UK's response to Argentina's invasion of the Falkland Islands, though Britain acted without the approval of the United Nations. However, this phrase does not cover a pre-emptive strike when a nation feels itself to be threatened by armed attack, as when Israel tried to justify its bombing of a nuclear reactor in Iraq as an act of self-defence. Attempts have been made to justify invasion as protecting the country's nationals on a number of occasions, for instance by UK and France in invading Suez in 1956. Justification of the actions in these cases can hardly be seen as valid: the action taken, even if necessary, was not proportionate and the reasons given were inadequate.[6]

Self-defence has also been used to justify action under Article 51 in response to terrorist activities. For instance, in 1986 the USA, with the support of the UK, bombed targets in Tripoli, justifying its action as a response to past terrorist attacks on nationals and as a deterrent to further such attacks in the future. In the Security Council Russia and the UK supported the US legal argument, China condemned it, and other states were non-committal.

A further problem has been brought to the fore by the use of 'pre-emptive self-defence'. The USA initially made it seem as though its attack on Afghanistan was directed against those responsible for the attacks on the Twin Towers and the Pentagon on September 11th 2001. In its September 2002 National Security Strategy the USA claimed the right to act alone to exercise its right of self-defence by acting pre-emptively against terrorists.[7] In October 2002 the US Congress used both the defence of the United States and enforcement of UN Security Council Resolutions to authorize the use of military force against Iraq. Significantly the UK and Australia, who also contributed forces to the initial attack, used only inferred authorization by the Security Council. The supposed threat from weapons of mass destruction subsequently proved to be without foundation, and attempts were made to justify the invasion publicly by claiming that it would lead to the overthrow of President Saddam Hussein and the replacement of a repressive totalitarian regime by democracy. The attack was seen by many to be illegal, but the UN had no means to stop it. The UN Secretary General stated that the view that states could, unilaterally or in coalitions, go to war 'represents a fundamental challenge to the principles on which, however imperfectly, world peace and stability have rested for the last fifty-eight years'.[8] In general, the use of pre-emptive self-defence as a justification for military action remains controversial and problematic.

A third major problem with the UN Charter is that it was written at a time when the urgent need was to prevent wars between states, and the increasing incidence of intra-state wars was not foreseen. In fact, the great majority of recent wars have been civil wars, but the Charter does not license the UN 'to intervene in matters which are essentially within the domestic jurisdiction of any state' (Article 2:7) unless the Security Council determines that there exists 'a threat to the peace, breach of the peace or act of aggression' (Article 39). Since the war in the Congo this has been used to establish the legality of intervention in a number of intra-state wars in Africa.

International law as enshrined in treaties has been extended by 'customary law', generated by the actions of states. Thus the International Court of Justice decided (1986) that the principles enshrined in the UN Charter had, by state practice, become international law. The basic principles that aggression is illegal, that one state should not forcibly intervene in the affairs of another, and that self-defence is permissible are part of customary law.

In summary, war is both morally and legally wrong unless authorized by the United Nations, though there may be rare occasions in which it can be justified on the grounds of self-defence.

INTERNATIONAL COOPERATION

Moral issues have become increasingly salient on the international level in another way. Just as our moral codes maintain a balance between selfish assertiveness and prosociality in dealings between individuals, so on the international level governments must consider the balance between inter-state competition and cooperation. The organization of the world into nation states puts the emphasis on the former, but the accelerating pace of globalization brings an increasing need for international cooperation.

Elected politicians are likely to do their best to satisfy the perceived needs of their electorate (see Chapter 7), but only the wise amongst them understand that their electorate is part of the human race, and that it is morally right to consider also the needs of others. And not only morally right: the increasing interdependence of nation states consequent upon globalization may make it pragmatic not to take a purely nationalistic perspective.

This last point has become vital with the rise of terrorism. Terrorism is totally unacceptable and must be countered in every way possible. But caution in the use of violent means is essential, because violence breeds violence in an enduring spiral. One must also try to understand the terrorists' behaviour. As has been pointed out by many thoughtful observers, the roots of terrorism are diverse, but lie in part in jealousy of the absurdly high standard of living of many Westerners, abhorrence of what is perceived as their immoral lifestyle, and the willingness of some Western countries to support undemocratic regimes when it suits their interests. It is reasonable to suppose that, if the money spent by the UK and the USA on the Second Gulf War had been judiciously spent on aid and education, the world would have been a more peaceful and happier place today. This is not just being wise after the event: many were urging such a course before the Second Gulf War and before Bush launched his so-called 'war on terror'.

Cooperation is not only a matter of sending aid to starving Third World countries, for it may mean entering agreements that disadvantage your own industry or agriculture. More importantly, it means supporting the United Nations, though perhaps a rejuvenated United Nations,[9] as arbitrator over international affairs, and supporting international agreements designed to ensure a more peaceful world.

The issue has become an urgent one in recent years because the US administration, seeing the USA as easily the most powerful nation in the world and able to disregard the needs of others, has withdrawn from, or failed to cooperate over, a number of international agreements intended to ensure a more peaceful world. For instance, the USA has refused to cooperate over measures designed to alleviate global warming; it refused to ratify the Comprehensive Test Ban Treaty; it refused to sign the Protocol to the Biological Weapons Convention on the grounds that the industrial secrecy of American firms might be affected; it refused to sign a treaty to abolish landmines; it refused to subscribe to the International Criminal Court unless its own nationals were exempted; it has posed difficulties for agreement on the control of small arms; and so on. In all these cases the USA is putting the short-term interests of its own nationals above steps towards a more equitable world. In the past, other nations, including the

UK, have used their power in their own interests in a manner which is now seen as inexcusable. We must learn from experience.

Of course, one must not forget the dilemma in which politicians are placed. Re-election matters to them, and they must seek to satisfy their electorate. But, instead of using the media to convince the public that their policies were devised solely in their interests, another course is open for them. In the long run it would be wiser to argue for the importance of a safer and better world.

THE CONDUCT OF WAR: GOVERNMENTAL DECISIONS

In the nineteenth century the act of declaring war freed a state from many of the restrictions to which it would otherwise have been subject. A state going to war was not restricted in the degree of force used, and could, for instance, impose a blockade on the enemy's coast. Once engaged in war, however, other laws, concerned for instance with the treatment of prisoners, came into effect. Since, presumably for pragmatic reasons, actual declarations of war were rare, whether or not two states were at war was often controversial.

However, changes in international law have led to the rules governing the conduct of war being based almost entirely on humanitarian, that is moral, considerations. As the international lawyer Christopher Greenwood writes:

The laws of war have become the laws of armed conflict, applicable whenever fighting takes place between states. . . . Above all, the rationale for those rules has changed and become the protection of basic human rights in times of armed conflict.[10]

Humanitarian issues had, of course, been recognized much earlier, but were seldom formalized in law. An early exception, protecting combatants, was the St Petersburg Declaration of 1868 outlawing explosive and inflammatory bullets on the grounds that they caused suffering beyond that necessary to achieve the military objective of disabling enemy soldiers. More recently the Geneva Conventions of 1949 and the Additional Protocols of 1977 deal with such matters as the treatment of prisoners and of civilians in the power of the enemy. They can therefore often be seen as being in opposition to the Principle of Military Necessity, which contributed to the Hague Rules of 1907 and governed, for instance, the powers of an occupying army over the territory it occupied.

The bases of the Geneva Conventions and Protocols are generally humanitarian, for instance requiring a commander to evaluate whether the probable military advantages of an intended plan of action justify the probable civilian casualties. Where military and humanitarian considerations conflict, it has been held that military necessity does not contravene a prohibition imposed by law.

With hindsight, the German Baedeker raids on British cities, and the heavy bombing of German and Japanese cities by Allied air forces in World War II, are cases in which supposed military necessity was allowed to predominate over humanitarian considerations. Whether those responsible for the bombing strategy had clearly thought through the issues and faced the terrible consequences of the mass bombing of civilians is an open question. Certainly, at any rate initially, the raids were presented to the public as involving targets of military importance (see below), and inertia made it difficult to abandon a policy once embarked upon.

The culmination of those raids came with the dropping of nuclear weapons on Hiroshima and Nagasaki.

THE USE OF NUCLEAR WEAPONS

The debate between military necessity, international law, and humanitarian considerations is nowhere more salient than in the case of nuclear weapons, and this case deserves special mention. How far should the scientists who invented the bomb be held responsible, and how far the politicians who ordered it to be used against Japan? Many of the scientists who developed these weapons were motivated by the concern that Nazi Germany was likely to develop them, and that the only way to stop them using them would be to threaten retaliation if they did. Yet when the defeat of Germany was imminent, and it was apparent to the Allied authorities that the Germans would be unable to make an atomic weapon, only one scientist resigned from the project— Joseph Rotblat. Regarded as a traitor and possibly a spy, he returned to the UK with undeserved ignominy. Rotblat devoted the rest of his life to research on the medical effects of radiation and campaigning for the abolition of nuclear weapons, and was eventually awarded the Nobel Peace Prize jointly with the Pugwash Conferences on Science and World Affairs, an organization in which he had played a leading role.[11] No doubt, many of the other scientists involved agonized over the political decision to use the bomb on cities, and some suffered acute pangs of conscience for the rest of their lives.

After the defeat of Nazi Germany, it was the politicians who took the decision to use nuclear weapons against Japan. Their use against Japanese cities was justified to the public on the grounds that it would bring World War II to an end and obviate the necessity for an invasion of Japan which might have involved very heavy American casualties. That was clearly an immoral decision, implying that thousands of Japanese civilian lives were worth less than the lives of American soldiers that might be lost if mainland Japan had been invaded. In fact there were covert political reasons for the dropping of the bombs related to ambivalence towards the USSR.

Given the test bombs that had already been exploded, and given the concerns expressed by a small proportion of the scientists involved, it is improbable that those responsible for ordering the use of the bombs were ignorant of the effects that they would have. In fact the two bombs dropped on Hiroshima and Nagasaki were responsible for more than 120,000 deaths immediately and many more from radiation sickness over the years. The Hiroshima bomb was more than a thousand times more powerful than the largest conventional weapon then available.

Some years later, the development of an even more fearsome device, the thermonuclear bomb, became a possibility. A committee of the US Atomic Energy Commission, chaired by Robert Oppenheimer, took a unanimous stand against its development. Truman decided otherwise, and Oppenheimer was branded as a security risk to the USA and ostracized. Other nuclear scientists, including Enrico Fermi, took a similar position, and called on the president not to proceed. Their advice was scorned. On the other side of the Iron Curtain Andrei Sakharov, at considerable personal cost, took a similar stand. Again, with the division between East and West deepening, his plea was ignored.[12]

Now we all know with certainty what nuclear weapons can do, there can be no excuse for their use, or for the threat to use them. Their inherent immorality is widely recognized, not only on the grounds of the immense suffering that they can cause, but also because they would almost always breach the principle of proportionality between the damage inflicted and the intended objective that was inherent in the Just War doctrine. Their illegality has been formulated in a number of international treaties including especially the Non-Proliferation Treaty (NPT), which came into force in 1970, and now has 188 signatories. In accordance with the treaty, all non-nuclear states that have signed it have undertaken not to acquire nuclear weapons.[13] At the same time, the five states that were officially recognized as possessing nuclear weapons by virtue of the fact that they had tested them by a certain date (the USA, Russia, France, China, and the UK)—have undertaken to get rid of these weapons.[14] The relevant Article VI reads:

Each of the Parties to the Treaty undertakes to pursue negotiations in good faith on effective measures relating to cessation of the nuclear arms race at an early date and to nuclear disarmament, and on a Treaty on general and complete disarmament under strict and effective international control.

By signing and ratifying the NPT, the nuclear member states became legally committed to nuclear disarmament. However, the hawks in these states, in an attempt to retain nuclear weapons, utilized an ambiguity in Article VI, which made it appear that nuclear disarmament is linked with the achievement of general and complete disarmament. But the NPT Review Conference—an official part of the implementation of the NPT—at its session in 2000, removed this ambiguity in a statement issued by all five nuclear weapons states. It contains

an unequivocal undertaking by the nuclear-weapon states to accomplish the total elimination of their nuclear arsenals leading to nuclear disarmament to which all States Parties are committed under Article VI.

This should have made the situation perfectly clear. But not only have the nuclear weapon states taken only minimal steps to implement its obligations, but the USA has introduced new policies, which directly contravene these obligations. The policy announced in the New Nuclear Posture Review and in later statements, as well as the decisions to develop new nuclear warheads, implies the indefinite retention of nuclear weapons in direct contradiction to the undertaking under the NPT. Furthermore, the recalcitrance of the original nuclear weapon states provides other states with an excuse for acquiring them, and the protective policy of the USA towards Israel has made it impossible for the international community to act over that country's illicit possession of nuclear weapons.

Moreover, the Bush and Blair administrations seem to have managed to convince many that only a part of the NPT, the part that applies to the non-nuclear states, is valid, and that therefore states which violate it—as, at the time of writing, Iran stands accused of doing—must be punished for the transgression. The part concerning the obligation of the nuclear states to get rid of their nuclear weapons has been deliberately obfuscated.

The obligation of the nuclear weapon states to disarm is seldom mentioned, nor is the USA's intention to develop new types of nuclear weapons. Yet the governments of the nuclear weapon states are clearly aware of the dangers of nuclear weapons. Indeed they are constantly saying how dangerous nuclear weapons are and that they must not be allowed to fall into the hands of undesirable elements or rogue regimes. What these governments do not say is that these weapons are almost as dangerous in the possession of friendly nations. We are facing here a basic issue in which the ethical and legal aspects are intertwined. The use of nuclear weapons is seen by the great majority of people in the world as immoral, owing to their indiscriminate nature and unprecedented destructive power. While they exist, the most stringent precautions against their accidental explosion and against the dispersal of radioactive material resulting from a nearby conventional explosion, are essential—and precautions to the necessary standards are not always met, especially since the dissolution of the Soviet Union. Their possession—and therefore likely eventual use—is thus equally unacceptable, whether by 'rogue' or benevolent regimes.

The elimination of nuclear weapons has been the declared aim of the United Nations from the beginning, and resolutions to this effect are passed, year after year, by large majorities in the General Assembly. The International Court of Justice has ruled that the use or threat to use nuclear weapons is illegal (though leaving open the question whether it would be illegal to use them if the survival of the state were at stake). These resolutions are ignored by the nuclear weapon states, as are all attempts to discuss the issue by the organ set up for this purpose, the Conferences on Disarmament in Geneva. Robert McNamara characterized current US nuclear policy as 'immoral, illegal, militarily unnecessary and dreadfully dangerous'.

CHEMICAL AND BIOLOGICAL WEAPONS

Comparable comments could be made about the elimination of biological and chemical weapons. Biological weapons were prohibited by the Biological Weapons and Toxins Convention in 1972, but the Soviet Union had a clandestine programme until at least 1990. The USA refused to support a verification protocol in 2001, but has adopted a national system of controls of its own. The convention on Chemical Weapons came into force in 1997 and, unlike the Biological Weapons convention, carries provisions to verify that it is not being violated. However, many problems remain, not the least being that chemical and biological weapons do not always require large facilities for their manufacture, and can be used by small groups more easily than nuclear weapons.

THE COST OF WAR

Some will argue that national defence must be a first priority for governments, and that justifies expenditure on the armed forces. For many states, at least, that represents an outdated priority, for the possession of a powerful army is often merely a matter of national prestige. Seldom considered is the tremendous drain on national resources caused by militarism. The bulk of the expenditure on arms is spent by Western nations, nearly half by the USA, but the effects of militarism are most felt by some of the poorer nations in the world, such as Rwanda, Burundi, Eritrea, and the states that constituted the former Yugoslavia. Money spent on arms is money that could have been spent on health, education, and welfare.

In addition, militarism can have detrimental effects on the environment. In peacetime, vast areas are used for training—the US Department of Defense controls 25 million acres[15] of land. In wartime, the environmental damage can be vast, two outstanding examples being the use of defoliants by the USA in the Vietnam War and the igniting of the oil wells in the First Gulf War.

THE CONDUCT OF WAR: DECISIONS ON THE GROUND

The motivation of combatants in pre-industrial society was probably based primarily on individual aggressiveness accompanied by individual exhilaration and long-term expectation of gain. Such issues are much less important in modern war because, as societies became more complex, institutions played an increasing role in their organization. To exemplify the sense in which I am using the concept of 'institution', in British society Parliament is an institution, with the roles of prime minister, ministers, Members of Parliament, the voting public, and many others. Each incumbent at each level has certain rights by virtue of the office he holds, and also certain duties. War can be seen as an institution, with many constituent roles—politicians, generals, combatants, munition workers, transport workers, doctors, and so on. The incumbent of each role does what he does in large part because it is his duty to do it. The munition workers make arms because that is their duty, the transport workers convey the arms to the front because that is their duty, and so on. The tank commander advances on the enemy, and the bomber pilot destroys his objective, not because they are aggressive, but primarily because they are doing their duty in the institution of war. Other motivations enter in, such as loyalty to buddies and even hope of glory, but duty is a major issue.[16]

The willingness of combatants (and of incumbents of other roles in the institution of war) to do their duty is assisted by perception of the enemy as somehow belonging to a race apart. This stems from the distinction between in-group and out-group basic to the ethics of interpersonal relations (see Chapter 2). It is accentuated by the efforts of propagandists to portray the enemy as evil or subhuman. And this is not merely a device, for those in charge may themselves perceive the enemy in this way. The decision to drop the nuclear bombs on Japan must have involved a disregard for Japanese lives, and during and since the 2003 Gulf War little effort was made even to estimate enemy civilian casualties.

The military manuals of the major powers instruct their forces that the laws of war are to be observed in any armed conflict. It becomes their duty to abide by those laws. In addition, commanders in the field are given 'rules of engagement' which set out how far they should go in pursuing 'military necessity'. Nevertheless, conflicts between moral issues, or between moral and pragmatic issues, often arise. We are not concerned here with the conflicts inherent in any military action between duty or obeying orders and fear of death or injury. But performing one's duty is seen as essentially a moral matter, especially when the safety of comrades depends on it, and duty may involve killing. In addition, killing is sometimes a matter of self-protection: the situation is seen as one of 'kill or be killed'.[17] Reflection may then not precede the act but, later, the priority of the need to protect oneself is seen as justifying it.

The duty to kill involves sharpening the boundary between the in-group and the out-group and a distortion of a basic precept: killing out-group members is praised while killing in-group members is still forbidden. Killing for its own sake occurs rarely in modern war, though very occasionally, in hand-to-hand combat, individuals seem to go berserk and revel in the act of killing. When combatants kill unnecessarily, as in the massacre of villagers by US forces in the Vietnam War at Mai Lai, and comparable incidents in the Second Gulf War, their actions are condemned. Both morality and international law condemn the killing of non-combatants and of prisoners of war, though such actions may sometimes be seen by the perpetrators as dictated by military necessity. Similar considerations apply to the use of torture to extract information from prisoners of war. Many countries have resorted to torture in the past, the Second Gulf War providing recent examples.

Sometimes, especially in civil wars, revenge is the basis of a duty to kill. Grudges, perhaps perpetuated over generations, are seen as requiring retribution: the killings in Northern Ireland and the former Yugoslavia were motivated in part by revenge, itself deliberately fostered by the politicians. The antipathy between Greeks and Turks on Cyprus has been fostered in the same way.[18] But most often, killing is seen as necessary simply as a duty. Duty in the institution of war is thus elevated to the status of a moral precept in opposition to the prohibition on killing. And duty, in its turn, may be ascribed to the need to obey orders in the service of a higher goal, such as the preservation of the motherland.

Some inhibition against killing is nearly always present, but killing is easier when it is not seen as involving killing other human beings. As noted above, for that reason the enemy is often portrayed in propaganda as subhuman: a US World War I recruiting poster portrayed the enemy as a gorilla carrying off a white woman, and bore the caption 'Destroy this mad brute'.

Bomber crews or artillerymen can find it possible to neglect the consequences of what they are doing because they do not witness the immediate consequences of their actions. No apology is necessary for illustrating this with two individual examples from World War II, as they involve also several of the issues mentioned above.

The first concerns Hein Severloh, a *Wehrmacht* machine-gunner defending Omaha Beach as the Americans landed. Firing at about 600 yards' range, he believes that he probably killed around two thousand Allied soldiers as they landed. In an interview sixty years later he wept with remorse as he said: 'What should I have done? I thought I would never get out of there alive.... It's them or me, that's what I thought.' But at the time the distant Americans looked 'like ants', and it was not that which was still giving him recurring nightmares. Rather it was the moment when he shot with his rifle a young American who came running up the beach closer to him. To quote a newspaper interview: 'Mr Severloh still remembers the man's contorted expression. "It was only then I realised I had been killing people all the time", he said, "I still dream of that soldier now. I feel sick when I think about it" '.[19]

The second example of the ease with which it is possible to kill unseen enemies comes from accounts of bomber crews attacking German cities in World War II. P. Johnson, a bomber pilot, describing (though with considerable understatement) his part in attacking a city that had already been identified with flares dropped by the Pathfinder force, wrote:

Our own part in the fighting was quickly over...what we had to do was search for the coloured lights dropped by our own people, aim our bombs at them and get away.[20]

However, with time the same author came to have great doubts about what he was doing. The 'Objective' in their operational orders changed from something like 'to destroy an enemy factory' to euphemisms like 'to do maximum damage to an enemy industrial centre' and later 'to destroy an enemy city'. As the war progressed, though he continued to carry out his duties, he gradually became more doubtful of their utility and morality. His autobiography provides a moving testimony of his conflict between duty bravely carried out and growing moral doubts. Thus he records how, in his position as a junior and middle-ranking commander becoming aware of the consequences of mass bombing on civilians,

I tried to convince myself that, in my relatively junior capacity...I had no right to differ from the plans which were made by people much more experienced than myself. But I did not succeed.[21]

I never flinched from the certainty that it was always essential that for middle leadership, people like me with some responsibility and influence, to give complete loyalty to the orders we were given.[22]

So he continued to take part in raids in which he knew that hundreds or thousands of civilians would be killed. Ordered to lead in the bombing of a small German town, he knew that enemy civilian

casualties were bound to be high because the roofs of cellars and shelters would collapse with the heat and a weight of rubble that they could not carry.[23]

He knew little of the Laws of War, but

I did know that a Law is little use without an authority to enforce it. In this case the only authority was moral feeling.[24]

Thus,

I could see I was trapped. I was sure that what we were to do was not only wrong but stupid...I had to believe that the top brass, civil and military, thought this [Neverthless] the best way to win the war, but this time I couldn't say to myself that 'Oh well they know best'.[25]

After the war he wrote:

I have no doubt now that those actions have left a permanent stain on the long, and on the whole honourable, record of British arms. It is worse that the actions were stupid as well as cruel.[26]

And he does not excuse himself from this indictment.

CONCLUSION

For centuries people have felt moral revulsion at the killing and suffering involved in war, but such feelings have been suppressed by other considerations, and war has been seen as necessary or inevitable. Nevertheless, over the years the laws governing war have changed to bring them closer to moral humanitarian feelings. War is now seen as both wrong and illegal unless authorized by the United Nations, and there are legal as well as moral constraints on its conduct. However, wars still occur, and both legal and moral considerations are bypassed.

The ethics of war differ, though only in degree, in two important ways from those we subscribe to in everyday life. The first is the sharpness of the distinction between 'us' and 'them'. In our ordinary lives we tend to treat members of the groups to which we belong—football team, church congregation, compatriots, or what have you—differently from non-members. But attitudes to non-members are seldom so negative as those towards the enemy in wartime. In wartime such attitudes are exacerbated by government propaganda, military training, and so forth. We must rest our hopes on the effect of globalization in extending our view of the in-group. It is now inconceivable that inter-state wars will recur in most of Europe: perhaps that feeling will be extended globally.

The second difference lies in the importance of duty. We have seen that the incumbents of each role in the war machine do what they do in part because it is their duty to do so. Duty influences the actions of both politicians and commanders as well as the combatants themselves. For the politician who feels it his duty to initiate war, difficult ethical decisions are inevitable, and he may feel torn between duty and its certain consequences. An oft-cited example is the remark by Sir Edward Grey, who was British Foreign Secretary on the eve of World War I: 'The lamps are going out all over Europe; we shall not see them lit again in our lifetime.' Of course, not all politicians see the dilemma as clearly as did Grey.

Military commanders must face a similar dilemma. Every action that they order is almost certain to result in casualties to their own side as well as to the enemy. For both commander and combatant, killing is a duty that must override the moralities of civilian life. Nevertheless the data show that many soldiers feel a strong aversion to taking life, even that of an enemy. Some military training is directed specifically towards helping soldiers to overcome their ambivalence, and that ambivalence is a welcome augury for the future.

How does the individual politician who declares war, or the individual combatant who kills enemy soldiers and even civilians, come to terms with what he has done? I have no doubt that the concept of duty is of primary importance here. This is clearly illustrated in the bomber pilot's account of his feelings, summarized above. The duty to obey the orders of his superiors was firmly incorporated in his self-concept, and he felt it must take precedence over his scruples. The concept of duty had no doubt been part of his self-concept since childhood duty to parents, to the family, to his friends. And duty had been accentuated by his military training—'Theirs but to do and die'. Duty to kill is facilitated by dehumanization of the enemy. The German machine-gunner said the men he was firing at looked 'like ants': it was the horror of having killed a real human being face to face that lived on in his nightmares. Propaganda is used to denigrate the enemy, emphasizing their foreignness and their immorality, thereby accentuating the barrier between in-group and out-group. Of course, duty and group difference are not the only ways in which people come to terms with killing or having killed, they may push their memories into the unconscious: veterans are notorious for not talking about their experiences. Past actions may be attributed to fear ('It was either him or me') or revenge. But duty and group loyalty are probably primary, and both have been accentuated by societal conventions and myths.

This is not the place to discuss in any detail how the abolition of war could be achieved, but a few words are necessary. The tragedy is that attempts to outlaw war by both the League of Nations and the United Nations in legal terms have had little success—primarily because states, and especially powerful states, have allowed their perceived self-interest to override their legal obligations. As discussed elsewhere,[27] there are many difficulties in reforming the UN, but if that were possible and it had more powerful sanctions at its disposal the UN could be the basis of a better world. It is the only supranational forum that we have, and therefore must be supported.

With modern communications, the public are becoming more aware of the horror of war. At the time of writing the apparently insoluble problem of the violence in Iraq is causing public revulsion. One hopes that public awareness of what war is really like will make politcians less willing to use force. There is a balance to be achieved here, for too much exposure can deaden the effect.

On the other hand, military censorship of what can be shown and the gradual diminution in the publicity given to the dozens of civilians being killed daily in Iraq is surely to be regretted.

Yet another route, and in the long term perhaps the most hopeful, is to attempt to eliminate the causes of war. No war has a single cause, it is always a matter of interaction between a number of causes. Competition for resources and for territory, poverty, differences in religion and ethnicity, the personal ambitions of politicians, and many other factors all play a role. A more equitable distribution of the world's resources within and between countries and improvements in education could reduce the effectiveness of many of the causes of wars, and would be important steps in the right direction.

But two things are essential for war and require special emphasis. One is weapons to fight them with. We must seek to abolish the arms trade, and so far as is possible to rid the world of virtually all weapons, and especially to eliminate weapons of mass destruction. Some will say that that is an unachievable aim, since almost anything can be used as a weapon. But there can be no doubt that the export of firearms from the more to the less industrialized states has facilitated violent conflict. The second essential factor is willingness by young men and women to use those weapons. As we have seen, the motivation of combatants and others in modern war is largely one of duty in the institution of war, so we must ask what can be done to remove or neutralize the factors that support the institution of war. These involve three categories of factors. The first includes a number of everyday issues, such as the use of warisms in everyday speech, the teaching of history as a history of wars, male chauvinism, and so on: these seem like trivial issues, but they help to make war seem an acceptable way of solving conflicts in the public mind. The second group, which has a similar effect, includes a number of pervasive cultural factors, such as national traditions of pride in military achievements, and the virtue of nationalism (in the sense of the denigration of other countries, not to be confused with patriotism, meaning love of one's own). The third, perhaps the most powerful of all, is the military-industrial-scientific complex responsible for the manufacture of weapons, but acting also as a major force in maintaining the respectability of war and inherently stable because of the career ambitions of the incumbents of its various roles.[28]

Perhaps the most important issue is not to lose faith in the possibility of abolishing war. It will not happen overnight, but every step is progress made. One must remember the tenacity of the opponents of slavery and of the suffragettes that eventually brought them to their goals. The appreciation of the common humanity of the people of every nation, and of every group within each nation, is the strongest basis for believing that one day in the not too distant future war will be seen to have been an immoral, aberrant, and bizarre way of settling disputes.

10

What Does All This Mean for the Future?

In the early chapters of this book I sketched an approach to understanding the nature of morality that involved asking how moral intuitions develop in individuals, what causes people to decide to act ethically, how morality evolved, and how it functions in the life of individuals and society. As a heuristic device, I focused on two basic propensities, prosociality and selfish assertiveness: both are the product of natural and cultural selection. Prosociality, roughly wanting to do good to at least some others, is as much part of being human as selfish assertiveness. I suggested that this approach leads to a view of morality rather different from that resulting from reflection on what people ought to do. In this final chapter I suggest why that is so and what it must mean to society.

In our everyday lives, ethical conflict is often present. Indeed some conflict will *always* be with us. However carefully we plan our society, propensities for both prosociality and selfish assertiveness will be there. Early in our evolutionary history natural selection operated both to promote success in competition with in-group members, and to promote prosociality towards them: inevitably these two propensities will often be in conflict. Much of the time we are not conscious of any conflict, because our conscience keeps us on the 'right' track. I have used the model of the 'self-concept' or 'self-system' to illustrate this: most of the time one acts according to the precepts and values incorporated into one's self-system, but feels guilt if one sees oneself to be behaving in a way that is incongruent with them.

If moral conflict is so common and so inescapable, is it inevitable that our lives should always be fraught with tension? Life in modern societies often seems like that, and it is not entirely due to the phrenetic activity induced by excessive materialism. But the tension is not so great as it might be. In Chapters 4 to 9, concerned with different contexts of modern life, I have emphasized not only the conflicts that each context inevitably entails, but also the devices that have been used by individuals and by society to reduce the tension that comes with mild departures from the moral rules of the society.

But how can that be? Surely, to be effective moral precepts must be perceived as absolute? Precepts that one can ignore at will would be useless. Here one must distinguish what is good for the society from what is good for the individual. For the integrity and well-being of society, most of its members must treat the rules as absolute. But the members of the society or group will compete as well as cooperate with each other. So perhaps some let-out from the rules is necessary: those who stick strictly to the rules could be outcompeted by those who bend them. Perhaps also, as conflict is so common, it is necessary to minimize the tension that results from minor deviations from the moral rules. Over minor issues in personal relationships tension can be reduced by the mechanisms that maintain congruence in the self-system (pp. 24–7). One can reinterpret or misperceive or re-evaluate one's own behaviour or re-evaluate the opinions of one's critics so that any lack of congruency is removed. What is important for tension reduction is that one should *perceive* the conflict to have been settled or one's own behaviour to have been correct. By maintaining congruency in the self-system, an individual can perceive his actions to be right when in fact his behaviour departs to a limited degree from the precepts generally accepted in the society.

Adjusting one's perceptions to suit one's conscience is a personal matter. A more important problem comes from the development of institutions within the society. If institutions, or the incumbents of roles within them, bend the rules too far, social disintegration could occur. We have found that some constraints are already in place. In business, the economists tell us that capitalism with its attendant competitiveness is good for the consumer, but some limits, moral and legal, have already been imposed on what is seen as legitimate competition. In spite of that, insufficiently constrained competition has led to unacceptable differences between rich and poor. Whether that is inevitable in a capitalist society is an open issue. In international conflict, attempts have been made to ensure that the means used to wage war are constrained within limits. We have internationally accepted Rules of War and more specific Rules of Engagement—but Hanover, Dresden, Hiroshima, Nagasaki, Guantanamo, and Abu Graib show that they are inadequate. (Of course it will be better when waging war by any means is simply not acceptable.) Again, the limits of political necessity should be set by the electorate. But are they?

We must not treat politics or business or war, or for that matter relationships, science, medicine, or law, as games with rules of their own, in the way that the rugby union player recognizes that those who play rugby league or American football have different rules. We must be constantly on our guard to see that the rules are not bent too far.

Rule-bending to suit the needs of institutions occurs in at last three ways.

First, the perceived boundary between the in-group and the out-group may be emphasized or changed. The most obvious example is war, where it becomes not only possible but meritorious to kill members of the out-group just because they are members of the out-group. The leaders of a group at war do their utmost to emphasize differences between *us* and *them* by propaganda, national anthems, or exploiting differences previously not seen as divisive, such as religion. Sometimes revenge for past injustices by the other group is seen to justify killing. In wartime the modification of the precept forbidding killing is usually accepted by all but a few members of the public—sometimes even by normally pacifist religious groups. A 'war mentality' becomes ubiquitous in part because the enemy are seen as dangerous non-persons.

War is not the only context in which the boundaries are changed or hardened. In peacetime, politicians have loyalties to country, party, trade unions, and general public, and these may conflict with each other as well as with their personal integrity: the boundaries of the in-group change with the context. In business, likewise, the executive is confronted with conflicting loyalties to shareholders, employees, customers, and so on. Duty to the in-group to which loyalty is owed justifies behaviour that might not otherwise be acceptable, but is passed off as 'business ethics'.

Second, moral duty to the in-group can often be seen as transcending other moral precepts. This is clearly another major issue in war. Duty (to the country, commander or comrades) not only causes combatants bravely to expose themselves to mortal danger but also justifies killing, disregarding the suffering of others, and sidestepping the Golden Rule. Indeed, not only is such behaviour accepted by society, the combatant is honoured for it. In business duty may lie in loyalty to the firm; in politics to the party; in law and medicine it is loyalty to the profession. In these cases duty to the in-group provides the actor with justification for departing from or stretching the everyday moral rules: whether it is seen to do so by outsiders is largely determined by their own biases. In science, perceiving scientists as an in-group some-times fosters research, but 'duty to science' has also been used to absolve scientists from considering the possible applications of their work. In medicine loyalty to the profession can foster moral rectitude.

Third, precepts may be distorted to accord with practice. In business the Golden Rule may become changed to Do-the-best-for-yourself-because-you-know-the-other-is-trying-to-do-the-best-for-himself: this receives further validation from the economist's view that competition is good for the consumer. In politics 'political necessity' may be used to justify the means used to gain the goal. In these cases, competitive business dealings and politicians representing parties are seen as necessary for the society, and the distorted precepts are at least partially accepted by others. The legal system, necessary to preserve the integrity of society, involves barristers defending individuals they believe to be guilty or prosecuting those they believe to be innocent, and judges and the prison service doing to others what they would not like to be done to them: this is accepted as necessary by the vast majority of the population.

More than one of these devices for distorting moral precepts may be used together. In wartime, governments use propaganda both to validate feelings of duty to the state and to exaggerate the differences between compatriots and the enemy. All who participate in the war effort are thereby enabled to feel that they are doing the right thing.

Furthermore, the route taken may differ according to the con-text. Consider again the case of killing, rigidly forbidden in our culture. There are some who follow the precept to the letter. They are pacifists in wartime, prosecuted in law for following their consciences; they oppose capital punishment; they may even be vegetarians on grounds of conscience. But the majority are more selective: they will say that killing is wrong, but are willing to go to war on grounds of duty. The basic source of that duty may be eradicating an evil, or the feeling that their country needs them. Some people condone capital punishment, perhaps on the grounds of a duty to protect society or to take just revenge. Religious belief has served as a pretext for murder: adherents to nearly all the world's major religions have seen it as their duty to kill those of a different faith at one time or another. In some cultures it is the duty of a man to kill a female relative because she has broken a sexual taboo: the culture dictates a duty that overrides the prohibition on killing.

Infanticide, regarded with horror in Western societies, has come to be seen as acceptable in some local cultures where poverty is extreme and the chances of rearing the child slender or, in other cultures, simply because the child is the wrong sex. Again, some see killing in the heat of jealous passion as forgivable or at least understandable because the murderer could not have controlled himself.

Most people condone accidental killing, or at least see it as a lesser crime than deliberate killing. However, there is a large grey area here: should the person responsible have been taking more care? Experimental evidence suggests that people see it as wrong to kill another deliberately as an intended means to save a larger number of others, but as acceptable when killing is a foreseen but unintended consequence of an attempt to save a greater number.[1] (The difference may lie in the fact that deliberately killing one individual breaks the Golden Rule, but saving several and incidentally killing one is not seen as doing so.) In general, then, while every culture prohibits killing, people may justify killing in a variety of ways according to the context, and the justifications are accepted by others in the society.

And now comes the problem. One can see this moral flexibility as backsliding, as due to lack of moral integrity on the part of individuals. Such a view is inescapable if one sees the precepts as inviolable. But perhaps sticking strictly to the rules is not the best strategy, for individuals who stuck strictly to the rules might easily be exploited by selfishly assertive individuals who disregarded them. Selfishly assertive individuals would come to predominate in the group, and the group would either self-destruct or be at a disadvantage with respect to other groups (pp. 36–7). Perhaps in human history the optimum balance between prosociality and selfish assertiveness required just a modicum of ability to bend the rules. Feeling guilty over major infringements of the moral code is necessary to preserve prosociality and cooperation in the group, but the availability of ways to maintain congruency in the self-sytem can assuage guilt over minor deviations (pp. 25–7). Such a mechanism could preserve individuals with mainly prosocial inclinations by enabling them to bend the rules in minor ways without suffering from guilt and without losing out in competition with free-riders.

Beyond that, the complexity of the societies that we have created virtually *requires* some ability to bend the rules. Given the ubiquity of conflict, it is difficult to see how a society with absolutely inflexible rules could persist in a changing world. The competitiveness inherent in the business world and in politics, patriotism and nationalism in wartime, loyalty to the profession, make it almost necessary for individuals to behave in ways that they might not otherwise adopt. In war, this is necessary to achieve national or group goals. In business, economists justify competition because it reduces prices for the consumer. One can even speculate that, early in human history, some flexibility in morals as well as in behaviour was necessary to cope with changing environments and climatic conditions.

In such cases, not only must individuals be preserved from guilt, but their behaviour must be condoned by society because the institution is seen to be beneficial to the society. That is where the soldier's duty, political necessity, business ethics, and the barrister's façade come in. The public recognizes its need for the soldier to do his duty, for the barrister to plead cases in which he does not fully believe, and feels it is inevitable, even though undesirable, that the politician be two-faced, or that the businessman should do the best for himself or the group he represents. That does not mean that public acceptance is absolute. In the cases of the soldier and the barrister it may be, but businessmen are treated with caution and, according to a recent poll, politicians are seen as among the least trustworthy of professionals.

But if flexibility were too great, social disintegration would ensue. In the extreme case, there would be no morality and no society, for the balance between prosociality and selfish assertiveness must be maintained if social life is to be possible. That is where legal systems come in. Originating in accepted morality, a major function of the law is to ensure that flexibility is limited and appropriate. Most of the law deals not with everyday misdemeanours but with the more extreme infringements of moral precepts. It is perceived as having no flexibility—though in practice societies have mechanisms for changing the law. But law itself could not exist if the lawmakers did not treat the lawbreakers in ways in which they would not want to be treated themselves.

In summary, one is confronted with a series of anomalies. Moral precepts must be seen as absolute. Yet minor infringements that lead to small adjustments to maintain congruence in the self-system may be generally accepted in the group: 'after all no one is perfect'. However, society sometimes demands that some individuals should bend the rules even further: behaviour that might be considered immoral in other circumstances must be condoned for the sake of the society. This results in a change, or a strengthening, of the perceived boundaries between in-group and out-group; a change in the relative status of the precepts or the duties that circumstances require; and/or distortion of the usual rules in order to accommodate the behaviour that the situation requires. But if this flexibility were too great or were used inappropriately, society would break down, so it must be limited by a legal system, also perceived as rigid and unmodifiable. Yet to be effective, the legal system itself must evolve and at any one time involves departures from morality as it is normally perceived, departures that are necessary if the law is to perform its function in society.

Morality is thus a different sort of thing from what I, and at least some others, possibly most others, had perceived. Rather than a set of rules by which one should try to abide, it must be seen as a delicately balanced system whose rules must be seen as rigid but can nevertheless be bent, but only to a limited extent and in special circumstances when it is necessary for personal or societal reasons. Bending the rules to a degree is necessitated by the society in which we live, given that human nature is what it is.

None of this is to be taken as justifying any old behaviour to gain our ends. Backsliding is not to be condoned. That people bend the rules, and that in certain situations society condones their behaviour, are facts: I would merely speculate that this situation, while perhaps not what we would like when seen from the outside, is largely a consequence of the society we have created for ourselves to live in. I am not advocating a completely relativist position: deviations condoned by society are specific to particular situations. The rules by which we try to live our lives have been constructed over time by humans to make social life possible. We must come to terms with their real nature without losing an iota of our respect for them.

The question remains, where are the pressures of modern social life taking us? There will always be some who want to bend the rules too far: how can we put on the brake?

WHAT CAN BE DONE?

Sometimes the world situation seems beyond repair, with the only possible course being to sit on one's hands and muddle along. That, of course, is a doctrine of despair, but the question of what can be done to make things better is not an easy one. Certainly just telling people they should be good is insufficient. One reason for that is that morality as a set of rules carved in stone gives a false picture of what morality is about. The world changes and moral codes change as it does so. In addition the increasing complexity of society demands that, in some contexts, the rules must have some flexibility for the sake of the society as a whole. Few will deny that, if we are going to make the world a better place, we *need* moral guidelines, but moral guidelines are more likely to command acceptance if their bases are understood and moral conflicts are more easily solved if their nature is recognized.

At first sight, the present approach may seem to have little to offer: if what is morally right is what is held to be right by 'society', it may seem that one has no option but to go along with the herd whatever one's own view may be, or to convince oneself that one's own view is the right one. That would be simplistic.

The view of morality advocated here does not bring ready-made solutions. Nor does it offer a set of precepts very different from those that most of us accept. But perhaps it offers some (interrelated) guidelines. First, it is initially a biological view, concerned with what people share as human beings. This is not only a matter of material needs, but includes the needs for constructive relationships and a fruitful social life. Shared needs provide the best starting point when cultures, moralities, or world views clash. The best route to 'understanding' between people brought up in different traditions, or who hold different world views, starts from an emphasis on the universal needs and deeds of human beings. We all eat, drink, and sleep. We all need understanding relationships. We all have systems of moral precepts to which the Golden Rule is basic. We each form a view of what life is about. It does not matter that we have different religions, myths, or social systems. Because we are all human beings, there is much in common across the diverse cultural arrangements that we have built. If we remember our humanity and acknowledge the similarities, the Golden Rule provides a route to understanding how different cultural practices and how even different religions can, in their different ways, satisfy common human needs. Similarly, in disagreements between individuals, progress can be made if each sees himself as the other, understanding the other's point of view. Revenge for past perceived injustice then becomes less salient as a possible course of action.

However, there is a danger here. As I have frequently stressed, what is natural is not necessarily good: sometimes we must resist the morality that natural selection has imposed on us. By nature we tend to limit our prosociality to an in-group. The criteria by which the in-group was originally defined depended largely on familiarity, though they could be adjusted socially. We tend to look after ourselves, and to confine our good deeds to our neighbours. Now the modern world demands that we extend the in-group, and to do that we must overcome our nature. For a more peaceful and equitable world, we must dilute the concept of national sovereignty and extend our in-groups to embrace humanity, including people who at first sight are far from familiar. In sending aid to countries devastated by famine, earthquake, or tsunami we are recognizing the global nature of humanity.

Here it is legitimate to ask by what criteria do I assert that we should be better off in some areas if we amended the morality that comes with our basic human nature. The answer is at least partly pragmatic: with more devastating weapons, war has become even less acceptable; and with globalization the different peoples of the world have become increasingly interdependent. The answer is also partly a moral one: with increasing knowledge of other peoples we can no longer blind ourselves to the fact that superficial markers like skin colour, language, customs, place of birth conceal an underlying humanity.

Second, although this approach draws on the principles of natural selection: they are not enough. In the real world moral precepts are conveyed largely through personal relationships, and what is purveyed is the product of a dynamic interchange between what people do and what they are supposed to do. While the bases of our moral codes lie in the forces of natural selection, the precepts we live by have been honed over the generations so that they fit (more or less) the context in which we live. Two things follow from this. One is that, other things being equal, our system of precepts is therefore likely to be the best guide for us. In everyday affairs, there has to be an extraordinarily good reason for bending the rules. But we must also remember that the precepts of other societies have been adjusted for social and physical circumstances that differ from our own, so that the dialectical relations between what people do and what they are supposed to do have resulted in rules that may differ in some ways from our own. Thus the precepts of other cultures must not be seen as *necessarily* wrong.

Third, circumstances change, the physical and social circumstances in which people live are not constant, and the rate of change seems to be increasing. Social life is complicated, even loving relationships can involve conflict, and the world in which we live is being transformed. So simple and unchangeable rules that will give straightforward guidance in any situation may give a false sense of righteousness. To understand the relations between rules and circumstances, we need to understand where the rules come from and their nature. For example, rules that concern matters that are closest to our biological roots, such as the Golden

Rule and those governing some aspects of mother–child relations, are pan-cultural. Societies without norms or rules concerning these issues are unlikely to succeed. The basic elements of sexuality are also pan-cultural, and very difficult to change: hence the great difficulty in eradicating prostitution. However, the precepts governing sexual behaviour, seen as rigid in most societies (though no longer quite so rigid in our own), differ between societies.

Again, a hierarchical structure is virtually pan-cultural: given human selfish assertiveness, that may be inevitable. If this leads to authoritarian rule, those at the bottom will suffer and they may rebel. A hierarchical structure can be stable only if social conventions cause the leader to pay for his privileges by services to those under him.

Fourth, with the increasing complexity of society, it becomes desirable, and sometimes necessary, that some people in some contexts should not follow rigidly rules intended to apply to all contexts. The institutions that maintain our society may require behaviour and values that, in other contexts, we might not see as desirable. The fact that moral precepts have some limited flexibility, though anathema to anyone who seeks for absolutes, places more responsibility on the individual and provides both a danger and a challenge. The ease with which we can perceive ourselves to be acting morally when we are not calls for constant vigilance.

Therefore, at the personal level, we must ask, 'Am I feeling virtuous because I am maintaining congruency in my self-system though I am actually bending the rules?' At the societal level, we must ask whether behaviour that infringes the general rules in particular contexts is justified. For example, the economists tell us that capitalism with its attendant competitiveness is good for the consumer (though not for everyone). However, our propensity for selfish assertiveness easily leads to lack of restraint. Unrestrained capitalism has already led to unacceptable differences between rich and poor. Whether that is inevitable in a capitalist society is an open issue. In the UK, for a few years after World War II, the gap between rich and poor was narrowed, but in the seventies and eighties it increased again. It seemed that the welfare state, which merely ameliorated the lot of the poor, was not enough. If the difference between rich and poor continues to increase one can be sure that new excuses or euphemisms extending beyond 'business ethics' will be created to make the situation seem acceptable. The notion that wealth would 'trickle down' was one of them. We must constantly be on our guard for others. As another example, in a world where disputes between states are often settled by war, it is in the interests of each society that combatants should kill those fighting on the other side. Surely this means that we must do our best to find ways of settling disputes other than violent conflict? Again, can it possibly be in the general good that politicians be licensed to lie?

So where does this lead in the long run? We must value the system of morality that we have, yet at the same time not seek to shelter under a set of absolutes. Flexibility does not mean everyone for themselves: if it did, families and societies would disintegrate. Some flexibility in moral rules may be necessary, but it must be limited. An outside limit is set by the pan-cultural Golden Rule, but everyday dealings require something more specific than that. The only way to ensure proper limits on flexibility is a proper understanding of where moral principles and precepts come from, and how they function to maintain a harmonious society. Where particular social contexts seem to demand flexibility in the rules, we must make sure that that is not the result of the self-interest of those involved, and that a better system is not possible.

Selfish assertiveness could become increasingly salient. A recent survey indicates a cultural shift towards selfishness: for instance the sales of high-emission cars has increased from one in eighteen cars ten years ago to one in eight today. The trend is evident in so many ways, from the increase in thefts from churches to the buying by schoolchildren of model answers to exam questions on the internet.[2] Competitiveness is an inevitable part of human nature, but it sometimes seems as though it is taking over our lives. In the commercial world, in sport, in the world of television, satisfaction no longer comes from doing well, only from doing better than others. Enjoying another's company takes second place to being stronger, more skilful, more clever, more verbal than the next man. And much of the time it is competition for money. Every second television programme seems to involve competition for money—and often also public shame on the losers. Professional football is no longer a game, it is a competition for money.

That means constant vigilance. We must watch carefully the development of societal institutions, because they may be fuelled by excessive selfish assertiveness. We must not be afraid to question the way our society is heading, but for two reasons we must do so hesitantly: first, it is a society that has worked so far, and second, every society involves a complex of interrelated parts, and the consequences of an apparently minor change may be difficult to predict.

Going beyond the present approach, there are two further steps to be taken. One is the obvious one of minimizing the causes of moral conflict. To a large extent, bending the rules is a consequence of the society in which we live. We may not be able to put the clock back but we can strive to minimize the ills of the world. If the world's goods were distributed more equitably, there would be less robbery; if there were no weapons, violence would be reduced; proper education could enhance mutual understanding. Such issues are well known: sadly so because, though often rehearsed, progress is slow. Moves to ameliorate the lot of the less well-off are barely holding their own. In the UK a national health service has replaced the old panel practice, we have old age pensions, we have social services—none perfect, perhaps, but steps in the right direction. Education in all parts of the world improves, helping people to maintain a balance between loyalties to society and their own autonomy, and to see beyond their own society to the global whole. We must strive to accelerate these trends.

The second step, more intangible than any mentioned so far, yet embracing them all and more important than any of them, looks to the future. We must try to steer society by creating a new world view.[3] This sounds like mushy idealism, and certainly a new world view will not be achieved overnight. But without idealism, we could easily slide backwards. In my opinion there is no alternative. The way forward must lie not with the improbable deities of religious fundamentalism, but by facing the realities of the world and of our nature. We must seek a world view that is based on acceptance of the interconnectedness of all people, even of all living beings. In particular, while continuing to value and perpetuate local cultures, we must learn to devalue the boundaries of nation states and value individuals from all cultures equally. As noted above, in this context what is natural is not what is right: we must overcome our natural predisposition to divide the world into in-group and out-group.

We need a world view that emphasizes our potentialities for prosociality and devalues the competitiveness that has become of overriding importance in modern societies. This is a matter of balance. Self-assertiveness and competitiveness will always be with us, but so also will prosociality. We can seek to emphasize the pleasure of a job well done. Let the choir value and enjoy together the music it produces, whether or not it wins the Eisteddfod; the team value a game well played, whether or not it wins; the scientist a problem solved, whether or not she got there first. Examples are to be found in contexts in which competitiveness is downplayed. In marathon running, the winner inevitably gets a prize, but there is a medal for every competitor who completes the course in a reasonable time. Climbing a mountain peak is a cause for satisfaction, never mind how long it took. How many Monroes[4] one has climbed may be a matter for pleasant competitiveness in the pub, but the real pleasure comes from associating with others who enjoy the mountains. The doctor takes pride in successfully treating his patients, not in comparing his successes with other doctors. Let us ask how far competition is really necessary. Competition may be a useful tool to promote the best choral singing, the more skilful player, ability in research, but just because selfish assertiveness is always with us, there is no need to worship winning.

This will sound impracticable and excessively optimistic to many, and will take more than one generation to achieve. But the morality of a culture both affects and is affected by how people behave, so what we do now matters. Creating a world conducive to prosociality means action. It is too easy to sit and hope that things will get better. Priority must be given to education, and especially education for parenthood. In a democracy everyone can try to change things. Non-governmental organizations, like Amnesty, Greenpeace, and various conservation societies are not only making a material difference, they are changing the way people see the world. In a democracy, the government depends on the electorate. You can write to your Member of Parliament or representative: your letter may not be read by the person to whom it is addressed, but it will be counted. Public demonstrations have an effect: the peace marches in the UK about the invasion of Iraq did not stop the war, but they did make a permanent difference to the climate of opinion. Local social groups of almost any kind—sports clubs, religious groups, professional societies, music societies, trade unions—can increase the health of a society, its so-called social capital. And perhaps most important are the little things, the way in which we behave to our neighbours every day.

We must be unashamedly optimistic. Certainty of purpose must be married to humility and tolerance. To many, it seems that war can never be abolished, yet there are parts of the world in which violent conflict formerly seemed endemic and now seems unthinkable: France and Germany and most of the European Union are obvious examples. There are even signs, faint it must be admitted, that the hegemony of the USA might become slightly more muted as a consequence of the horrors of the Second Gulf War. Wealth differentials between states have been increasing, yet the richer states are beginning to recognize their responsibilities to the poorer ones.

And there are other hopeful signs. I have already mentioned the increasing (though still inadequate) willingness to send aid beyond national borders. Racism is on the retreat, although experiencing temporary setbacks as a consequence of the activities of terrorists. Racist remarks are simply no longer tolerated in the majority of countries. Religious differences still pose problems, but at least doctrinal differences between Catholics and Protestants are an issue only when exploited by politicians. Although exacerbated by fundamentalists, the dispute in Palestine is maintained primarily by the question of landownership rather than the religious difference. The importance of education is generally recognized, and illiteracy is decreasing. There will be setbacks, like the failure in every country except

Finland of the 1974 UNESCO agreement to promote Education for Peace in all schools, but there are signs that the time is now ripe for further initiatives of the same sort.

No one individual can create the world we want, but every individual can make a difference. And there is no other course.

Appendix
Relations to Moral Philosophy

I have claimed that this is a relatively new approach to under-standing morality.[1] Of course, it still stands on the shoulders of those who have gone before, but it is a set of shoulders different from those usually associated with discussions of morality. Per-haps, therefore, I should spell out how this approach both differs from and resembles some of the traditions in moral philosophy. It is perfectly reasonable for you, the reader, to ask whether a new approach could possibly add anything to the deliberations of philosophers over hundreds, indeed thousands, of years. How-ever, Bernard Williams, writing about the limitations of modern philosophy, is a little dubious about where moral philosophy has got to, in part because 'The resources of most modern moral philosophy are not well adjusted to the modern world'.[2] I have in fact been surprised to find how far the present approach accords in some ways with Williams's vision for the future of moral philosophy, though it departs from it in others. It would be premature to claim that it does any better, though I would not have written this book if I did not have my own views on that. But it does lead to a somewhat different view of the nature of morality, as I described in Chapter 10. I am no philosopher, and my reading in philosophy is limited: Professors Jane Heal and

Malcolm Schofield have done their best to help me to navigate round the generalizations that tempt an outsider, but bear no responsibility for the end product.

Starting point. While most philosophers focus on the individual as a rational agent, I have stressed the dynamic interaction between individual and the group, an interaction that implies limited lability in moral precepts. For many moral philosophers the central focus is on how people ought to behave; mine is on how they *think* they should behave and how they *actually* behave.

Goals. To an outsider, the goals of moral philosophers seem diverse. Some seem to be seeking for an abstract truth that lies outside the human condition, some for the logical status of moral claims of truth or falsity, some for a prescription for how an individual should live a good life, or for how an individual should act in particular circumstances; others discuss how an ideal society should be organized to maximize welfare or to minimize pain, or seek for the solution to a current problem based on the principles espoused by one or other of their predecessors. In addition, some philosophers attempt to build an ethical theory that can be tested. The current approach is seeking a theory only in the sense that natural selection is a theory: its validity depends on further evidence that confirms it. It seeks to understand how people come to think they ought to behave in some ways and not in others in real life situations in a particular society. That leads to an objective grounding for ethical beliefs in the historical and current circumstances of a society. It seeks not for a theory that embraces our diverse ethical intuitions, but to explain the origin of those intuitions. For present purposes only it includes, as well as behaviour recognized as morally good, conventions and obligations because one has a choice (at some level) whether or not to abide by them. While the early moral philosophers were concerned with improving the quality of individuals' lives, this approach is concerned primarily with why individuals act in this way rather than that.

Methods. The work of many, perhaps most, moral philosophers depends primarily on rational reflection that is general and abstract, although an increasing number are becoming concerned with the findings of modern science. The present approach is built upon the behavioural sciences—biology and ethology, psychology (developmental and social), sociology, and anthropology. In contrast to most philosophers, I have tried to use data that bear on the question of *why* people think they should behave in this way rather than that. Following ethological principles, I have sought four answers to the question *why* in terms of development (how morality develops in individuals); causation (what causes people to behave morally); what is the function of morality; and how has it evolved.[3] Chapters 1 to 4 summarize the answers to these interrelated questions. The view of morality as having evolved in a particular way and as constantly developing is part of what leads to a view of morality different from some traditional ones.

Of the four answers to the question why, the discussion of development draws on a large and growing body of evidence from developmental psychology (pp. 17–23). The suggested scheme for the biological evolution of morality is supported by modelling approaches and cross-cultural studies (Chapter 2). That for the more recent cultural evolution of morality is more speculative and involves piecing together fragments of evidence from legal history and other sources (Chapters 1 and 3). These questions of biological and cultural evolution stand together with the proposals for the functions of morality. The qualitative model of how morality affects behaviour is perhaps the least well substantiated: the concept of the self-system is currently receiving a good deal of attention in psychology (pp. 24–7): it is useful at the behavioural level but bypasses some questions of causation, for instance that of how precepts are stored.

Stability and change. During periods of stability, what an individual sees as right to do or value in a given society is usually what is seen as right by the majority, or by an influential minority, in the group in question.[4] This is compatible with contextualist approaches to morality, with the Aristotelian view that individual welfare is bound up with respect from other group members, and with Wittgenstein's claim of shared understanding based in social practice. It is also compatible with Williams's concept of confidence, since similarity with others helps validate the rules (pp. 46–8). It is also in accord with empirical data on how nonmoral behaviour not only involves negative feelings in the actor, but also indignation from others.[5]

While it is now generally recognized that morals change, not all philosophers are concerned with the mechanism of change, Hume and Nietzche being notable exceptions. The present account uses considerations from both biological and cultural evolution. The role ascribed to natural selection does not imply a Rousseauesque equation between what is natural and what is good. Nor does it imply continuous improvement in some abstract sense, as claimed by J. S. Huxley and some biologists. But it does recognize that we still carry characteristics that evolved under natural selection in quite different environments from those in which we now live. Most students of human evolution now agree that, although natural selection acts through individual reproductive success, that can be facilitated by harmonious group-living: hence it may pay to be prosocial to fellow group members.

Criticisms of an evolutionary approach by philosophers on the grounds that it is concerned with biological fitness and not well-being[6] overlook the complex relations between the two, and ignore the evidence that prosociality can be the product of selection.

Cultural selection, often neglected in attempts to take a biological approach, permits more rapid change than natural selection. Moral precepts are created, maintained, and decay largely as the result of two-way interactions between what people do and what they are supposed to do. In addition, influential individuals or groups may impose values or precepts that are conducive to their own well-being, and individuals may come to accept the precepts as moral. Such precepts are not necessarily conducive to the welfare of individual or group.

Human Nature. In spite of the dangerous nature of the concept, both this approach and that of moral philosophers must take into account human nature, though the characteristics on which they focus tend to differ. Aristotle saw our moral judgements as rooted in human nature. Hume emphasized the roots of our moral talk in our responses and feelings about each other. In discussing human needs, philosophers focus on such things as liberty, wealth, and well-being: in considering the role of natural selection this approach includes selection for characteristics conducive to successful survival and reproduction.

Many philosophers postulate a 'rational actor' who is interested in his own well-being but also has a propensity to behave prosocially. The present approach recognizes that actors are not always rational, and emphasizes the central importance of prosociality for individuals living in a viable society.

I have made a distinction between features that are common to all cultures (human nature in a strict sense) and those common to members of a particular culture (the loose sense).

Two potentials. Nietzsche places a good deal of his emphasis on the 'bad' side of human nature, while many philosophers are primarily interested in 'good' behaviour, seen against a background of 'bad' behaviour. I have argued that it is useful to think in terms of two similar but usually opposing potentials, for prosociality and selfish assertiveness. Selfish assertiveness includes behaviour selected initially to foster success in competition with other members of the group; prosociality behaviour selected for contributions to facilitating good relations within the group (which earlier may have led to success of the group in competition with other groups). Developmentally, 'good' and 'bad' behaviour are two sides of the same coin, which predominates depending largely on the vicissitudes of experience.

Limiting discussion to these two categories is, of course, a crude heuristic device: how we behave is a lot more complicated than that. I make no attempt to reduce motivations to one or two basic types such as altruism or pleasure-seeking, and my use of the concepts of prosociality and selfish assertiveness is not to be interpreted in that way. Data from developmental psychology show that humans are born neither 'trailing clouds of glory', later to be perverted by the evils of the world, nor intrinsically evil and selfish, needing to be taught how to behave. They have potentials for both. Furthermore I am not saying that selfish assertiveness is necessarily bad and prosociality good, for a degree of selfish assertiveness is necessary for many prosocial actions, and virtues that we see as positive can be put to antisocial ends.

Thus this approach sees an individual as balancing his own needs against what is good for relations in the group to which he sees himself as belonging. But much of what he sees himself as needing will benefit the group: this includes good relationships with others. At the same time this approach recognizes the complexity of human motivations, pro- and antisocial, and of the consequences of the ongoing interactions between experiential influences, currently held beliefs about ethical behaviour, and how people actually behave. How far it will help to untangle that complexity remains to be seen.

Absolutes. I distinguish between *principles* which are pan-cultural, such as the Golden Rule of Do-as-you-would-be-done-by, and *precepts*, which may be culture-specific. This renders unnecessary any discussion as to whether moral rules in general are absolute or relative: the principles are absolute, the precepts relative. There is common ground here with those Greek sophists who accepted that justice differs between states and argued that inside the 'conventional man', who followed the rules of his particular state, lay a 'natural man', who would be at home anywhere.

The Golden Rule has long been implicit in philosophical discussions: for instance, as stresed in Chapter 1, it is intrinsic to the approaches of Hume, Kant, and Rawls. In my view the Golden Rule is not a principle reached by rational reflection, but a matter of empirical fact, and I suggest that its basis lies in the importance of reciprocal exchange in human relationships. Unless an individual's behaviour towards another is initially prosocial, it is unlikely to be reciprocated positively. The basis of reciprocal exchange is in turn to be found both in studies of human development, and in studies of cultural and biological evolution. Reciprocity is perhaps related inversely to the importance of revenge in keeping order in early human groups.

Most moral precepts are statements of, or at least compatible with, the Golden Rule for specific situations—don't kill because you would not like to be killed. Indeed it is difficult to see that social life could be possible without some variant of the Golden Rule. Thus the partial answer to the Platonic question 'What is it about an action that permits us to call it just?' is not that it results from Divine guidance or from 'moral intuition', but 'It conforms to the Golden Rule'.

Moral precepts also differ between communities within a society. Members of one group may properly judge the code of another by the degree to which it conforms to the Golden Rule, but not by the extent to which the precepts differ from their own without understanding any differences in history and circumstances. Subgroups, such as gangs and terrorist groups, can elaborate their own precepts and values. Even individuals may have their own views: in this way, the rules can be bent. In addition, several chapters in this book have shown how, in certain contexts, bent rules can be accepted as inevitable or normal by other members of the society.

What is seen as right may differ with the situation, so there is no single moral answer and there are no general objective tests for morality except that most actions, to be moral, must be conducive to group harmony. The answer to the philosopher's question of how one should live can never be a simple one: it may differ between societies, between groups, and even between individuals. Rational discussion has its place in fostering change or stability, not in a search for the Absolute or for the essence of justice. On these issues this approach is compatible with the views of philosophers such as Hegel and Williams.

Ubiquity. Many philosophers imply that one should behave morally to everyone, or at least to all who come into the category of 'persons'. By contrast, the evolutionary approach indicates that, initially, prosociality was selected to be directed only to members of the same group. In the complexity of the modern world, most people belong to many groups, some as large as a nation state. Each group uses a variety of devices to increase the loyalty of its members. While most moral philosophers are right (at least for pragmatic reasons) to say that we should be as keen to help suffering people on the other side of the world as those in our own backyard, the fact is that we are not. The reason for that lies in our history, or prehistory. For pragmatic reasons there is a growing need for loyalty to be directed to all humankind: prosociality directed universally will require an adjustment of moral perspective.

Intentionality. While the usual view is that moral behaviour must be intentional, a result of the individual's will, I make no sharp distinction between prosocial behaviour that is spontaneous, resulting from precepts previously incorporated (by nature or nurture) into the self-system, and that which stems from conscious deliberation. Actions that appear spontaneous often depend on precepts that have been incorporated into the self-system in the past. Reason may have played no part either in the original acquisition of precepts or their deployment in a particular situation. In any case, many decisions are between conflicting spontaneously recognized moral oughts.

On this issue there is a symmetry between doing good and doing wrong. An unintended good act is often seen as less praiseworthy than an intended one, an unintended wrong as less deserving of blame than an intended one.

However, this approach does not necessarily align us with intuitionist philosophers. We have seen that seeking for congruency in our self-system may lead us to deceive ourselves into believing that what we are doing is right when our peers would say it is not. Leaving that aside, this approach raises the question, which I do not discuss, of how far asking what we perceive to be right is 'really' right is a sensible question to ask.

Is and Ought. This scientific approach can be seen as arguing either from what 'is', namely morality as we find it, or from current 'oughts', to the nature of 'oughts'. The hope is that, given an understanding of the nature of oughts, oughts appropriate to a current problem can more easily be identified. Some may say that this involves deducing what ought to be done from what is done, and thus infringing the conclusion of Hume and many others that it is logically impossible for facts to provide a basis for morality. Even Einstein wrote, 'Scientific statements of facts ... cannot produce ethical directives'.[7] The linguistic basis of this is that *description* cannot lead to *prescription* about what to do. But I am not concerned with logical deduction from what is to how one ought to behave, but with using what has been and is to surmise what is likely to be a good way to act now.[8] If it is considered that I have crossed this forbidden frontier, the question still remains, is that necessarily a heinous crime? The important issue is to be clear about the nature of the argument.

My argument is that the current 'oughts' of a given society have been honed over time by interaction between what people do and what they are supposed to do, and provide a basis for the precepts incorporated in their self-systems. Because morality is shaped by the forces of natural and cultural selection, and with certain reservations, these precepts provide a reasonable approximation to the best way to maintain the well-being of individuals within the group. A scientific approach can take us a long way in understanding morality and in analysing moral problems in all walks of life because it seeks not to find the essence of goodness or to use rational argument to determine what is or is not 'right' in general, but to seek for the sources of the ethical judgements we make in real life. It helps us to understand why we perceive this to be right and not that.

Moral decisions. Moral decisions may be either spontaneous, or the result of lengthy consideration. Thus neither 'reasoning' nor 'emotions' are adequate in themselves. And reasoning may be influenced by emotion or by preceding biases.

Role of religion. Although this approach postulates no absolute external authority for moral codes, it does not imply that religion has made no contribution to morality. Christianity, for instance, has purveyed moral precepts across the centuries, though neither Christianity nor Judaism can properly claim to have originated their basic principles.[9]

NOTES

INTRODUCTION

1. There are diverse views over the distinction between morals and ethics (e.g. Williams, 1985). I have treated them as more or less equivalent, but use 'morality' more as a collective for societal rules and 'ethics' for rules applicable to particular situations.
2. I have used 'he' and 'she' interchangeably, unless the context demands otherwise.
3. Williams, 1985.
4. Alexander, 1987; Darwin, 1901; Huxley, 1966. See discussion by Flew, 1974.
5. Boyd and Richerson, 2005. See p. 36–40 in this book.
6. Turiel, 1998.
7. Tinbergen, 1963.

CHAPTER 1

1. Some will see this as a Marxian emphasis on the difference between moral ideals and real life. I am trying to demonstrate the nature of that difference in different spheres of life and to specify its nature, though with no conscious Marxian bias.
2. Contrast Moore, 1903.
3. The discussion in this chapter is based on a more detailed one in Hinde (2002). Many of the references given there are not repeated here. Many of the issues discussed qualitatively here are treated quantitively in the papers of Boyd and Richerson, especially in Boyd and Richerson, 2005.
4. MacIntyre, 1967.
5. For more detailed discussion of the bases and role of religious beliefs, see Boyer, 1994, 2002; Hinde, 1999.
6. I use culture to refer to ways in which human groups or sub-groups differ that are communicated between individuals. Culture

is viewed as existing in the minds of the individuals in a society, and as in a continuous process of creation, maintenance, and decay through the dialectical relations with the behaviour of individuals (see pp. 12–13). The customs, beliefs, values, etc. of individuals are interrelated: the parts and their interrelations are referred to here as the 'socio-cultural structure'. As a concept it is related to the 'social imaginary', which refers to the ways in which people 'imagine their social existence ... the expectations which are normally met, and the deeper normative notions and images which underlie these expectations' (Taylor, 2004, p. 23: see also Pretorius, 2006).

7. McGrew, 2004; Byrne et al., 2004.

8. Brown, 2004; Tooby and Cosmides, 1992.

9. Cultural norms may involve antisocial as well as prosocial behaviour but similar principles of development apply in both cases. Thus mothers living in a society with aggressive norms may encourage selfish assertiveness in their children, For convenience, and admittedly loosely, I shall sometimes treat them similarly.

10. Hill and Hurtado, 1996.

11. Hsu, 1998.

12. Hinde, 2002; Lahti and Weinstein, 2005.

13. A view related to Hegel's dialectical change.

14. Nisan and Kohlberg, 1982; see review by Lahti and Weinstein, 2005.

15. Huxley, 1947.

16. Küng and Kuschel, 1993; Kohlberg 1981.

17. Kant, 1781. See also Rawls, discussed on pp. 258–9.

18. Appiah, 2006.

19. Harter, 1999.

20. Rheingold and Hay, 1980.

21. Kagan, 1989; Eisenberg and Fabes, 1998.

22. Baumrind, 1971; Bowlby, 1969/1982; Hinde et al., 1993.

23. Granqvist, 2006.

24. It is sometimes convenient to divide moral development into a series of stages: see e.g. Kohlberg, 1981.

25. When we start to inquire into the bases of moral rules, the answers we are given (or invent ourselves) are of two main types. Some claim that a moral rule must be followed because of its consequences, while others claim that moral rules are absolute and must be followed regardless of their consequences. These are often

distinguished as 'utilitarian' and 'deontological' respectively. On the present view, most moral precepts have become part of the culture because of their value for the well-being of the society. However, if one is interested in the immediate causes of behaviour, moral precepts are often followed simply because the individual 'knows' they are right, without thinking about the consequences. Thus adherence to a particular set of religious beliefs by the individuals in a society may augment the integrity of that society, but it is doubtful if the martyr being burnt at the stake has that in mind—we presume that he believes that following the First Commandment is a 'must'—though his resolution may be augmented by a belief that his steadfastness may ensure his place in an afterlife. In other cases, and especially those involving problems new to society, such as prenatal genetic screening (p. 136), decisions are made only after careful consideration, though the result may be heavily influenced by pre-existing biases.

26. Boyd and Richerson, 1985.
27. I use the term 'correct behaviour' here deliberately in view of the distinction made by Turiel (1998) between morals and conventions. Justifications for judgements about morals concern the welfare of others, while justification for judgements about conventions concern the social organization of the group in question. I have discussed this distinction elsewhere (Hinde, 2002, p. 8).
28. The 'self-concept' or 'self-system' is not to be thought of as a physical entity, but as a tool useful for understanding behaviour.
29. McGuire and McGuire, 1988.
30. Fincham and Bradley, 1989.
31. Backman, 1988.
32. Wright, 1984.
33. Hinde, 1987.

CHAPTER 2

1. e.g. Fried, 1978.
2. Williams, 1985, p. 197.
3. See Humphrey, 1997, for the relations between them.
4. Hamilton, 1964.

5. Trivers, 1974.
6. Trivers, 1985.
7. Trivers, 1974. The complexity of within-family conflicts has been worked out in some birds, e.g. Hinde and Kilner, 2007.
8. Daly and Wilson, 1996; Hrdy, 1999.
9. Silk, 1980, 1990.
10. Petrinovich, 1995.
11. Harcourt and de Waal, 1992.
12. Sober and Wilson, 1998.
13. Boyd and Richerson, 1991.
14. Alexander, 1987; Trivers, 1985.
15. The answer to the first is D and 7, and to the second 20 years and Beer.
16. Cosmides and Tooby, 1992.
17. Mealey et al., 1996.
18. Alexander, 1987, 2004; Boyd and Richerson, 2005.
19. Fehr and Gachter, 2002.
20. Boyd and Richerson, 2005.
21. Alexander, 1987; Dunbar, 1996, 2004; Novak and Sigmund, 1998.
22. Goffman, 1959.
23. Ekman and Friesen, 1975; Tooby and Cosmides, 1992.
24. Lahti and Weinstein, 2005.
25. Boyd and Richerson, 1992, 2005.
26. This assumes certain cognitive abilities, such as the ability to ascribe the actions of others to intentions.
27. Hawkes, 1993. See also E. L. Smith, 2004.
28. See discussion in the next chapter.
29. Milgram, 1974.
30. Harris, 1986.
31. Harman, 2000.
32. Ralls et al., 1988.
33. Hamilton, 1964; Low, 2000.
34. Goody, 2000.
35. Boyd and Richerson, 2005.
36. Rabbie, 1991; Tajfel and Turner, 1986.
37. Byrne et al., 1966.
38. Lahti and Weinstein, 2005.

CHAPTER 3

1. Independently, but entirely in keeping with the present thesis, Hoffman (2004) has proposed that three central principles are present in all societies and have parallels in law: (1) Promises to reciprocate must be kept; (2) Reciprocal exchanges must be relatively equal; and (3) Serious violations of the first two must be punished.
2. e.g. Woodburn, 1982; Black, 2000; Boehm, 2000.
3. Evans-Pritchard, 1940.
4. *New York Times*, 24 July 2006.
5. Bottéro, 1992.
6. Richerson and Boyd, 1999.
7. Adams, 1876.
8. Richerson and Boyd, 1999.
9. Saltman, 1985.
10. Hart, 1961.
11. Discussion in Hinde, 2002.
12. The concept of common property presumably has its origins in that of private property, but is often in conflict with it.
13. Certain groups have been in the past, or are, excluded: Native Americans, blacks, suspected terrorists.
14. In many societies, lawmakers have claimed divine guidance. Some laws, serving the interests of their makers, are therefore seen as moral. Loyalty to the leader might be an example. The promise of obedience to one's superiors in the Anglican Catechism is seen as a moral issue, though it may well have been included because it was in the priests' interests.
15. O'Hara, 2004; Greene and Cohen, 2004; Sapolsky, 2004.
16. A point made much earlier by Bentham, 1789; Devlin, 1958.
17. Larry Selk, personal communication.
18. In the USA, the North took notice when cotton mills were established in the South, making the South able to process its own cotton.
19. Devlin, 1965.
20. Warnock, 1998.
21. Turiel, 1998.
22. This chapter is not concerned with theoretical systems of justice. However, one system of justice which involves an attempt to balance the benefits between individual and community must be mentioned. Rawls (1971) has suggested that each individual should

have an equal right to the most extensive basic liberty compatible with a similar liberty for others; and that social and economic inequalities should be such as to be to everybody's advantage, and attached to positions and offices open to all. This amounts to a modified utilitarianism. Moral principles are binding because they would be accepted by a prudent individual designing a society when unaware what their own position in it would be.

23. An exception can be made if the accused has confessed his guilt.
24. Greene and Cohen, 2004; Sapolsky, 2004.
25. Hirsch and Ashworth, 1998.
26. Faulkner, 1994.
27. O'Hara and Yarn, 2002.
28. Box, 1983, 1992.
29. Daly and Wilson, 1996.
30. Jones, 2000, 2004.

CHAPTER 4

1. Brief review, Hinde, 1997.
2. Kagan, personal communication.
3. Stake, 2004. Exchange is commonplace in the grooming relationships of non-human primates, but the evidence for it in other contexts is scanty (McGrew, 2004).
4. Boyd and Richerson, 2005.
5. Kelley, 1979; Miller and Parks, 1982.
6. Lerner, 1974.
7. Prins et al., 1993.
8. Gintis, 2000.
9. Haley and Fessler, 2005.
10. Henrich et al., 2005.
11. Cosmides and Tooby, 1992.
12. Trivers, 1985; Ekman and Friesen, 1975.
13. Mealey et al., 1996.
14. Alexander, 1987.
15. O'Hara, 2004.
16. Alexander, 1987.
17. V. L. Smith, 2004.
18. Helwig, 1995.

19. Milgram, 1974.
20. Pearce and Littlejohn, 1997.
21. Oyserman et al., 2002.
22. Tinbergen, 1963.

CHAPTER 5

1. This chapter and the next overlap and should be read together.
2. The complex ethical problems that arise in the conduct of social science research are discussed by Oliver, 2003.
3. Founded 1660.
4. Recently, related academies for engineering and medicine have been established. See www.royalsoc.ac.uk/news.
5. Editorial, *Nature* (2006).
6. The emphasis on 'First past the Post' is partly due to editors of journals being unable to publish duplicated results. The internet could provide a solution to that.
7. Poison gas in World War I was an early example of using weapons of mass destruction.
8. Royal Society, 2004.
9. Langley, 2006; Parkinson, 2006.
10. Journé and Reppy, 2004; Petersen, 2004.
11. See Chapter 9, note 11.
12. Professor Sir Joseph Rotblat, who first drafted this chapter, was strongly in favour of oaths for scientists: the reservations are the present author's.
13. Hansen, 2005.
14. Royal Society (n.d.).
15. Coming from a scientist, some will see this as hypocrisy. I hope it is not. Certainly pride in their calling is to be promoted in young scientists.

CHAPTER 6

1. Journé and Reppy, 2004.
2. Meade, 2003.
3. Bennett and Duke, 1995.
4. Warnock, 1998.

5. *Guardian*, 21 April 2005.
6. Rees, 2003; Sokol and Bergson, 2005.
7. Asch, 2000; Walters and Palmer, 1997.
8. It has recently been claimed that a cell can be withdrawn from an early embryo without affecting the embryo's further development.
9. Tambiah, 2003.
10. A comprehensive discussion is given by Harris, 2004. See also Baltimore, 2005.
11. Somerville, 2000.
12. Royal Society, 2005.
13. In the latter case, it is believed that 'small quantities' was interpreted very liberally.
14. Nuffield Council on Bioethics, 2002, 2005; Weatherall, 2006.
15. Bateson, 2005.
16. These properly include regulations about the housing of animals. In some cases these are more restrictive than the scientific work needs, and result in unnatural conditions inimical to animal well-being.
17. Use of Non-Human Primates in Research. Weatherall, 2006.
18. Nuffield Council on Bioethics, 2005.
19. The Hippocratic oath was originally limited to a small group of practitioners, and was very different from what it is generally thought to be today. For instance, it did not permit surgery.
20. Other more basic issues may be involved, such as the rise of a consumerist attitude to medicine.
21. *Independent*, 11 March 2005.
22. Gunn and Masellis, 2005; Gunn et al. 2005.

CHAPTER 7

1. Bosher, 1965.
2. Dallek, 2004.
3. Tetlock, 2005.
4. Woodward, 2004, p. 296.
5. Ibid., p. 338.
6. Ibid.
7. Sen, 2004.
8. Hansard, 5 March 2004.

9. Statement by the NPT Review Conference, 2000.
10. Hansard, 5 March 2004.
11. Woodward, 2004, p. 129.
12. Papadakis, 1995.
13. Murray, 1995.
14. Carnegie Commission, 1997, p. 6.

CHAPTER 8

1. V. L. Smith, 2004, p. 76.
2. Sandel, 2005.
3. Homans, 1961, p. 76.
4. Lerner et al., 1976.
5. Frank, 2004.
6. Valentine and Fleischman, 2004.
7. Some will justifiably argue that 'loyalty' is used inappropriately when applied to a choice made for pragmatic reasons. There are interesting issues here concerned with the relation between motivation at one level of social complexity and consequences at another, but these cannot be pursued here.
8. In some contexts the government may itself act as a competitor, for instance in the provision of a postal service. It can also act as an employer, for instance in purchasing armaments.
9. Korukonda and Bathala, 2004.
10. Dasgupta, 2006.
11. G. Monbiot in the *Guardian*, 29 March 2005.
12. Box, 1983, 1992.
13. Cheung and King, 2004.
14. Michelson et al., 2004.
15. GlaxoSmithKline, The impact of Medicines; undated.
    Pfizer. Healthy business, healthy lives, healthy communities; undated.
16. Ishtar, 2003.
17. Carter, 2005.
18. Dasgupta and Mäler, 2000.
19. Amnesty, 131, 20.
20. Chryssides and Kaler (1996), who also provide illuminating discussions of many of the other issues discussed in this chapter.

21. Axinn et al., 2004.
22. *Independent*, 15 May 2004.
23. Tamagno and Aasland, 2000.
24. Gowri, 2004.
25. Zwetsloot and Marrewijk, 2004.
26. *Guardian* 28 November 2005.
27. Dasgupta, 2001.
28. Anon., 2006.
29. Matten and Moon, 2004.

CHAPTER 9

1. Speech to UN, 24 September 1961.
2. Wrangham and Wilson, 2004; Wrangham, in press.
3. Hinde and Rotblat, 2003.
4. Collier, 1991.
5. Many further examples are given by Gray, 2004.
6. Ibid., p. 129.
7. National Security Strategy of the United States, September 2002.
8. Secretary-General's address to the General Assembly, 23 September 2003.
9. Hinde and Rotblat, 2003.
10. Greenwood, 1991.
11. Braun et al., 2007.
12. Calogero, 1993.
13. Weeramantry, 2003.
14. Peacerights, 2004.
15. Parkinson and Webber, 2005.
16. Hinde and Rotblat, 2003.
17. Presumably an instance of the hypothesis that fear is a cause of aggression.
18. Papadakis, 1995.
19. *Independent*, 5 June 2004.
20. Johnson, 1995, p. 177.
21. Ibid., p. 203.
22. Ibid., p. 239.
23. Ibid., p. 242.
24. Ibid., p. 243.

25. Ibid., p. 245.
26. Ibid., p. 342.
27. Hinde and Rotblat, 2003, pp. 182–3.
28. Ibid., ch. 10.

CHAPTER 10

1. Petrinovich, 1995; Hauser, 2006.
2. S. Pemberton, 'Its all about me', *Guardian*, 8 July 2006.
3. Clark, 2002.
4. Scottish peaks over 3,000 feet above sea level.

APPENDIX

1. See Hinde, 2002 for more detailed discussion.
2. Williams, 1985, p. 197.
3. Tinbergen, 1963.
4. This implies nothing about individual preferences.
5. e.g. Hare, 1981.
6. Williams, 1985.
7. Einstein, 1950, p. 114.
8. Williams, 1985, p. 126.
9. MacIntyre, 1967.

REFERENCES

Adams, H. (1876). *Essays on Anglo-Saxon law*. Boston MA: Little, Brown.

Alexander, R.D. (1987). *The biology of moral systems*. New York: Aldine de Gruyter.

—— (2004). 'Evolutionary selection and the nature of humanity'. In V. Hosle and C. Illies (eds.), *Darwinism and philosophy*, pp. 424–95. Notre Dame IN: University of Notre Dame Press.

Anon. (2006). *Britain in Business*, 2, 22–4.

Appiah, K.A. (2006). *Cosmopolitanism: ethics in a world of strangers*. London: Allen Lane.

Asch, A. (2000) (Book review). 'The ethics of gene therapy'. *Politics and the Life Sciences*, 19, 291–2.

Axinn, C.N., Blair, M.E., Heoshiadi, A. and Thach, S.V. (2004). 'Comparing ideologies across cultures'. *J. Business Ethics*, 54, 103–19.

Backman, C.W. (1988). 'The self: a dialectical approach'. *Advances in Experimental Social Psychology*, 21, 229–60.

Baltimore, D. (2005). 'Limiting science: a biologist's perspective'. *Daedalus*, 134, 7–15.

Bateson, P. (2005). 'Ethics and behavioural biology'. *Advances in the Study of Behaviour*, 35, 211–33.

Baumrind, D. (1971). 'Current patterns of parental authority'. *Developmental Psychology Monographs*, 4 (1 and 2).

Bennett, L. and Duke, J. (1995). 'Research note: decision-making processes, ethical dilemmas and models of care in HIV/AIDS health care provision'. *Sociology of Health and Illness*, 17, 109–19.

Bentham, J. (1789). *Introduction to the principles of morals and legislation*. Edited W. Harrison. Oxford: Oxford University Press.

Black, D. (2000). 'On the origin of morality'. *J. Consciousness Studies*, 7, 107–19.

Boehm, C. (2000). Conflict and the origin of social control. *J. Consciousness Studies*, 7, 79–101.

Bosher, J.F. (1965). *The New Cambridge Modern History*, vol. 7, chap. 29, pp. 565–91. Cambridge: Cambridge University Press.

Bottéro, J. (1992). *Mesopotamia: writing, reasoning, and the gods*. Chicago IL: University of Chicago Press.

Bowlby, J. (1969/1982). *Attachment and loss*. vol. 1. *Attachment*. London: Hogarth Press.

Boyd, R. and Richerson, P. (1985). *Culture and the evolutionary process*. Chicago: University of Chicago Press.

—— —— (1991). 'Culture and cooperation'. In R.A. Hinde and J. Groebel (eds.), *Cooperation and prosocial behaviour*, pp. 27–48. Cambridge: Cambridge University Press.

—— —— (1992). 'Punishment allows evolution of cooperation (or anything else) in sizeable groups'. *Ethology and Sociobiology*, 13, 171–95.

—— —— (2005). *The origin and evolution of cultures*. Oxford: Oxford University Press.

Boyer, P. (1994). *The naturalness of religious ideas*. Berkeley CA: University of California Press.

—— (2002). *Religion explained*. London: Vintage.

Box, S. (1983, repub. 1992). *Power, crime and mystification*. London: Routledge.

Braun, R. et al. (2007). *Joseph Rotblat: visionary for peace*. Berlin: Wiley.

Brown, D.E. (2004). 'Human universals'. *Daedalus* (Fall 2004), 47–54.

Byrne, D., Nelson, D. and Reeves, K. (1966). 'Effects of consensual validation and invalidation on attraction as a function of verifiability'. *J. Experimental Social Psychology*, 2, 98–107.

Byrne, R.W., Barnard, P.J., Davidson, I., Janik, V.M., McGrew, W.C., Miklósi, A. and Wiessner, P. (2004). 'Understanding culture across species'. *Trends in Cognitive Science*, 8, 341–6.

Calogero, F. (1993). 'Responsibility of scientists and hopes for future peace in the world'. *Accademia Nazionale dei Lincei*, 104, 169–87.

Carnegie Commission (1997). *Preventing deadly conflict*. New York: Carnegie Corporation.

Carter, J. (2005). *Our endangered values*. New York: Simon and Schuster.

Cheung, T.S. and King, A.Y. (2004). 'Righteousness and profitableness: the moral choices of Confucian entrepreneurs'. *J. Business Ethics*, 54, 245–60.

Chryssides, G. and Kaler, J. (1996). *Essentials of Business Ethics*. London: McGraw Hill.

Clark, M.E. (2002). *In search of human nature*. London: Routledge.

Collier, J.G. (1991). 'Legal basis of the institution of war'. In R.A. Hinde (ed.), *The institution of war*, pp. 121–32. London: Macmillan.

Cosmides, L. and Tooby, J. (1992). 'Cognitive adaptations for social exchange'. In J.H. Barkow, L. Cosmides and J. Tooby (eds.), *The adapted mind*. New York: Oxford University Press.

Dallek, R. (2004). *Kennedy: an unfinished life*. London: Penguin.

Daly, M. and Wilson, M. (1996). 'Violence against stepchildren'. *Current Directions in Psychological Science*, 5, 77–81.

Darwin, C. (1901). *The descent of man*. London: Murray.

Dasgupta, P. (2001). *Human well-being and the natural environment*. Oxford: Oxford University Press.

_____ (2006). Lecture delivered to British Ecological Society. Oxford, 2006.

Dasgupta, P. and Mäler, K.-G. (2000). 'Net national product, wealth, and social well-being'. *Environment and Development Economics*, 5.

Devlin, P. (1958). *Morals and the criminal law*. Maccabaean Lecture, British Academy.

_____ (1965). The enforcement of morals. London: Oxford University Press.

Dunbar, R.I.M. (1996). *Grooming, gossip and the evolution of language*. London: Faber & Faber.

_____ (2004). *The human story*. London: Faber & Faber.

Ebadi, S. (2006). *Iran awakening*. London: Rider.

Editorial (2006). 'US to rule on research patent'. *Nature*, 440, 587.

Einstein, A. (1950). *Out of my later years*. London: Thames & Hudson.

Eisenberg, N. and Fabes, R.A. (1998). 'Prosocial development'. In W. Damon and N. Eisenberg (eds.), *Handbook of child psychology* (5th edn.), vol. 3, pp. 701–78. New York: Wiley.

Ekman, P. and Friesen, W.V. (1975). *Unmasking the face*. Englewood Cliffs NJ: Prentice-Hall.

Evans-Pritchard, E.E. (1940). *The Nuer*. Oxford: Clarendon Press.

Faulkner, D. (1994). 'Relational justice: a dynamic for reform'. In J. Burnside and N. Baker (eds.), *Relational justice*, pp. 159–174. Winchester: Burnside.

Fehr, E. and Gachter, S. (2002). 'Altruistic punishment in humans'. *Nature*, 415, 269–72.

Fincham, F.D. and Bradbury, T.N. (1989). 'Perceived responsibility for marital events'. *J. Marriage and the Family*, 51, 27–35.

Flew, A. (1974). 'Evolutionary Ethics'. In W.D. Hudson (ed.), *New studies in ethics*, vol. 2, pp. 217–86. London: Macmillan.

Frank, R.H. (2004). *What price the moral high ground?* Princeton NJ: Princeton University Press.

Fried, C. (1978). *Right and wrong*. Cambridge MA: Harvard University Press.

Gintis, H. (2000). *Game theory evolving*. Princeton: Princeton University Press.

Goffman, E. (1959). *The presentation of self in everyday life*. New York: Anchor Books.

Goody, J. (2000). *The European family*. Oxford: Blackwell.

Gowri, C. (2004). 'When responsibility cannot do it'. *J. Business Ethics*, 54, 33–50.

Granqvist, P. (2006). 'Religion as a by-product of evolved psychology'. In P. McNamara and E. Harris (eds.), *Where God and Science meet: how brain and evolutionary studies alter our understanding of religion*, vol. 2, pp. 105–150. Westport CT: Greenwood.

Gray, C. (2004). *International law and the use of force*. Oxford: Oxford University Press.

Greene, J. and Cohen, B. (2004). 'For the law, neuroscience changes nothing and everything'. *Phil. Trans. Roy. Soc. B*, 359, 1775–86.

Greenwood, C. (1991). 'In defence of the laws of war'. In R.A. Hinde (ed.), *The institution of war*, pp. 133–47. London: Macmillan.

Gunn, S.W.A., Mansourian, P.B., Davies, A.M., Piel, A., and Sayers, B. McA. (eds.) (2005). *Understanding the global dimension of health*. New York: Springer.

Gunn, S.W.A. and Masellis, M. (eds.) (2005). *Humanitarian medicine*. Chisholm: International Association for Humanitarian Medicine.

Haidt, J. and Joseph, C. (2004). 'Intuitive ethics'. *Daedalus* (Fall 2004), 55–66.

Haley, K.J. and Fessler, D.M.T. (2005). 'Nobody's watching? Subtle cues affect generosity in an anonymous economic game'. *Evolution and Human Behavior*, 26, 245–56.

Hamilton, W.D. (1964). 'The genetical evolution of social behaviour'. *J. Theoretical Biology*, 7, 1–92.

Hansen, T.B. (2005). *Teaching ethics to science and engineering students*. Copenhagen: Center for Philosophy, University of Copenhagen.

Harcourt, A.H. and De Waal, F.B.M. (eds.) (1992). *Coalitions and alliances in humans and other animals*. Oxford: Oxford University Press.

Hare, R.M. (1952). *The language of morals*. Oxford: Clarendon.

___ (1981). *Moral thinking*. New York: Oxford University Press.

Harman, G. (2000). *Explaining value*. Oxford: Clarendon.

Harris, J. (1986). 'The Survival Lottery'. In P. Singer (ed.), *Applied ethics*. Oxford: Oxford University Press.

___ (2004). *On cloning*. London: Routledge.

Hart, H.A.L. (1961). *The concept of law*. Oxford: Clarendon.

Harter, S. (1999). *The construction of the self*. New York: Guilford.

Hauser, M.D. (2006). *Moral Minds*. New York: HarperCollins.

Hawkes, K. (1993). 'Why hunter-gatherers work'. *Current Anthropology*, 34, 341–62.

Hegel, G.W.F. (1942). *Hegel's Philosophy of right*. Trans. T.M. Knox. London: Oxford University Press.

Helwig, C.C. (1995). ' "Adolescents" and young adults' conceptions of civil liberties'. *Child Development*, 66, 152–66.

Henrich, J., Boyd, R., Bowles, S. et al. (2005). ' "Economic man" in cross-cultural perspective: experiments in 15 small-scale societies'. *Brain and Behavioral Sciences*, 28, 795–855.

Hill, K. and Hurtado, M. (1996). *Aché life history: the ecology and demography of a foraging people*. Hawthorne NY: Aldine de Gruyter.

Hinde, C.A. and Kilner, R.M. (2007). 'Negotiations within the family over the supply of parental care'. *Proceedings of the Royal Society B*, 274, 1606, 53–60.

Hinde, R.A. (1987). *Individuals, relationships and culture*. Cambridge: Cambridge University Press.

___ (1997). *Relationships: a dialectical perspective*. Hove, Sussex: Psychology Press.

___ (1999). *Why Gods persist*. London: Routledge.

___ (2002). *Why good is good*. London: Routledge.

___ and Rotblat, J. (2003). *War no more*. London: Pluto.

___ Tamplin, A. and Barrett, J. (1993). 'Home correlates of aggression in preschool'. *Aggressive Behavior*, 19, 85–105.

Hirsch, A. von and Ashworth, A. (1998). *Principled sentencing: readings on theory and policy*. Portland OR: Hart.

Hoffman, M.B. (2004). 'The neuroeconomic path of the law'. *Phil. Trans. Roy. Soc. B*, 359, 1667–76.

References

Homans, G.C. (1961). *Social behavior: its elementary forms*. London: Routledge Kegan Paul.

Hrdy, S. (1999). *Mother Nature*. New York: Pantheon.

Hsu, E. (1998). 'Moso and Naxi: the house'. In H. Oppitz and E. Hsu (eds.), *Naxi and Moso ethnography: kin, rites, pictographs*, pp. 67–99. Zurich: Völkerstundemuseum.

Hume, D. (1896). *A treatise of human nature*, III. I, I. London: Oxford University Press.

Humphrey, N. (1997). 'Varieties of altruism—the common ground between them'. *Social Research*, 64, 199–209.

Huxley, J.S. (1966). *Essays of a humanist*. Harmondsworth: Penguin.

Huxley, T.H. (1947). 'Evolution and Ethics'. In J.S. and T.H. Huxley (eds.), *Evolution and ethics*. London: Pilot Press.

International Council for Science (2005). *Science and society: rights and responsibilities*. ICSU Strategic Review.

Ishtar, Z. dé (2003). 'Poisoned lives, contaminated lands'. *Seattle J. for Social Justice*, 2, 287–307 and other papers in this number.

Johnson, P. (1995). *Withered garland*. London: New European Publications.

Jones, O.D. (2000). 'On the nature of norms: biology, morality, and the disruption of order'. *Michigan Law Review*, 98, 2072–103.

—— (2004). 'Evolutionary analysis in law: an introduction and application to child abuse'. *North Carolina Law Review*, 75, 1117–242.

Journé, V. and Reppy, J. (2004). 'Report on Pugwash workshop on science, ethics and society'. *Pugwash Newsletter*, 41, 1, 3–11.

Kagan, J. (1989). *Unstable ideas: temperament, cognition and self*. Cambridge MA: Harvard University Press.

Kant, I. (1781). *Critique of pure reason*.

Kelley, H.H. (1979). *Personal relationships*. Hillsdale NJ: Erlbaum.

Kohlberg, L. (1981). *The philosophy of moral development*. New York: Harper Row.

Korukonda, A.R. and Bathala, C.R.T. (2004). 'Ethics, equity and social justice in the new social order'. *J Business Ethics*, 54, 1–15.

Küng, H. and Kuschel, H.J. (1993). *A global ethic*. London: SCM Press.

Lahti, D.C. and Weinstein, B.S. (2005). 'The better angels of our nature: group stability and the evolution of moral tension'. *Evolution and Human Behaviour*, 26, 47–63.

Langley, C. (2006). *Scientists or soldiers? Career choice, ethics and the military*. Folkestone, Kent: Scientists for Global Responsibility.

Lerner, M. (1974). 'Social psychology of justice and personal attraction'. In T.L. Huston (ed.), *Foundations of interpersonal attraction*. New York: Academic Press.

Lerner, M.J., Miller, D.R. and Holmes, J.G. (1976). 'Deserving and the emergence of different forms of justice'. *Advances in Experimental Social Psychology*, 9, 133–62.

Lin, E. and Lien, S. (2005). Manuscript, Judge Institute, Cambridge.

Low, B.S. (2000). *Why sex matters*. Princeton NJ: Princeton University Press.

Lyons, D. (1984). *Ethics and the rule of law*. Cambridge: Cambridge University Press.

MacIntyre, A. (1967). *A short history of ethics*. London: Routledge & Kegan Paul.

McGrew, W.C. (2004). *The cultured chimpanzee*. Cambridge: Cambridge University Press.

McGuire, W.J. and McGuire, C.V. (1988). 'Content and process in the experience of self'. *Advances in Experimental Social Psychology*, 21, 97–144.

Marx, K. (1963). *Early writings*. Ed T.B. Bottomore. London.

Matten, D. and Moon, J. (2004). 'CSR education in Europe'. *J. Business Ethics*, 54, 323–37.

Meade, T.W. (2003). 'Ethical and social issues arising from the science of medicine'. In *Meeting the challenges of the future*. Balzan Symposium, 2002, pp. 79–89. Florence: Olschki.

Mealey, M., Daood, C. and Krage, M. (1996). 'Enhanced memory for faces of cheaters'. *Ethology and Sociobiology*, 17, 119–28.

Michelson, G., Wailes, N., van der Laan, S. and Frost, G. (2004). 'Ethical investment processes and outcomes'. *J. Business Ethics*, 52, 1–10.

Milgram, S. (1974). *Obedience to authority*. New York: Harper Row.

Mill, J.S. (1863) *Utilitarianism*. London.

Miller, G.R. and Parks, M.R. (1982). 'Communication in dissolving relationships'. In S. Duck, (ed.), *Personal relationships*, 4. London: Sage.

Moore, G.E. (1903). *Principia ethica*. Cambridge: Cambridge University Press.

Murray, D. (1995). 'Families in conflict: Pervasive violence in Northern Ireland'. In R.A. Hinde and H. Watson (eds.), *War: a cruel necessity?*, pp. 68–79. London: Tauris.

Nisan, M. and Kohlberg, L. (1982). 'Universality and variation in moral judgement'. *Child Development*, 53, 865–76.

Novak, M.A. and Sigmund, L. (1998). 'Evolution of indirect reciprocity by image scoring'. *Nature*, 393, 573–7.

Nuffield Council on Bioethics (2002). *Genetics and human behaviour*. London: Nuffield Council on Bioethics.

——— (2005). *The ethics of research involving animals*. London: Nuffield Council on Bioethics.

O'Hara, E.A. (2004). 'How neuroscience might advance the law'. *Phil. Trans. Roy. Soc. B*, 359, 1677–84.

O'Hara, E.A. and Yarn, D. (2002). 'On apology and consilience'. *Washington Law Review*, 77, pp. 1121–92.

Oliver, P. (2003). *The student's guide to research ethics*. Maidenhead: Open University Press.

Oyserman, D., Coon, H.M. and Kemmelmeier, M. (2002). 'Rethinking individualism and collectivism'. *Psychological Bulletin*, 128, 3–72.

Papadakis, Y. (1995). 'Nationalist imaginings of war in Cyprus'. In R.A. Hinde and H. Watson (eds.), *War: a cruel necessity?*, pp. 54–67. London: Tauris.

Park, H. (2005). 'The role of idealism and relativism as dispositional characteristics in the socially responsible decision-making process'. *J. Business Ethics*, 56, 81–98.

Parkinson, S. (2006). *Corporations and career choice in science and technology*. Folkestone, Kent: Scientists for Global Responsibility.

Parkinson, S. and Webber, P. (2005). *Soldiers in the laboratory*. UK: Scientists for Global Responsibility.

Peacerights (2004). *Report of the inquiry into the legality of nuclear weapons, King's College, London*. London: Peacerights.

Pearce, W.B. and Littlejohn, S.W. (1997), *Moral conflict: when social worlds collide*. Thousand Oaks CA: Sage.

Petersen, A. (2004) 'Report from Pugwash Workshop on science, ethics and society'. *Pugwash Newsletter*, 41, 1, 11–13.

Petrinovich, L. (1995). *Human evolution, reproduction and morality*. New York: Plenum.

Pretorius, J. (2006). 'Defending the post-Apartheid State: How the RMA is informing the South African security imaginary'. Ph.D. thesis, Cambridge.

Prins, K.S., Buunk, B.P. and van Yperen, N.W. (1993). 'Equity, normative disapproval, and extra-marital relationships'. *Journal of Social and Personal Relationships*, 10, 39–63.

Rabbie, J.M. (1991). 'Determinants of instrumental intra-group cooperation'. In R.A. Hinde and J. Groebel (eds.), *Cooperation and prosocial behaviour*, pp. 238–62. Cambridge: Cambridge University Press.

Ralls, K., Ballou, J.D. and Templeton, A. (1988). 'Estimates of lethal equivalents and the cost of inbreeding in mammals'. *Conservation Biology*, 2, 185–93.

Rawls, J. (1971). *Theory of justice*. Cambridge MA: Harvard University Press.

Rees, M. (2003). *Our final century*. London: Arrow.

Rheingold, H. and Hay, D. (1980). 'Prosocial behavior of the very young'. In G.S. Stent (ed.), *Morality as a biological phenomenon*, pp. 93–108. Berkeley CA: University of California Press.

Richerson, P.J. and Boyd, R. (1999). 'Complex societies: the evolution of a crude superorganism'. *Human Nature*, 10, 253–90.

Royal Society (2004). *Nanoscience and nanotechnologies*. RS Policy Document 20/04. London: Royal Society.

——— (2005). *Personalised medicines*. RS Policy Document 18/05. London: Royal Society.

——— (2006). *Science in society*. London: Royal Society.

——— (n.d.). *Science and the public interest*. London: Royal Society.

Saltman, M. (1985). 'The law is an ass: an anthropological appraisal'. In J. Overing (ed.), *Reason and morality*, pp. 226–39. London: Tavistock.

Sandel, M.J. (2005). 'Markets, morals and civic life'. *Bulletin of the American Academy of Arts and Sciences*, 58, 6–15.

Sapolsky, R.M. (2004). 'The frontal cortex and the criminal justice system'. *Phil. Trans. Roy. Soc. B*, 359, 1787–96.

Sen, A. (2004). *On ethics and economics*. Oxford: Blackwell.

Silk, J. (1980). 'Adoption and kinship in Oceania'. *Amer. Anthrop.*, 82, 799–820.

——— (1990). 'Human adoption in evolutionary perspective'. *Human Nature*, 1, 25–52.

Smith, E.L. (2004). 'Why do good hunters have higher reproductive success?' *Human Nature*, 15, 343–64.

Smith, V.L. (2004). 'An economic perspective'. *Daedalus* (Fall 2004), pp. 67–76.

Sober, E. and Wilson, D.S. (1998). *Unto others*. Cambridge MA: Harvard University Press.

Sokol, D. and Bergson, G. (2005). *Medical ethics and law*. London: Trauma.

Somerville, M.A. (2000). 'Is human cloning inherently wrong?' In *Science for the twenty-first century*, pp. 209–10. UNESCO.

Stake, J.E. (2004). 'The property "instinct"'. *Phil. Trans. Roy, Soc. B*, 359, 1763–74.

Tajfel, H. and Turner, J. (1986). 'The social identity theory of intergroup behaviour'. In S. Worschel and W.G. Austin (eds.), *Psychology of intergroup relations*, pp. 7–24. Chicago IL: Nelson.

Tamagno, S. and Aasland, T. (eds.) (2000). *Invitation to a dialogue: corporate social responsibility*. Oslo: Norsk Hydro.

Tambiah, S. (2003). 'The stem cell research discussion in the USA'. In W. Ruegg (ed.), *Meeting the challenges of the future*. Balzan Symposium, 2002. pp. 113–22. Florence: Olschki.

Taylor, C. (2004). *Modern social imaginaries*, pp. 47–54. London: Duke University Press.

Tetlock, P.E. (2005). *Expert political judgement*. Princeton NJ: Princeton University Press.

Tinbergen, N. (1963). 'On aims and methods of ethology'. *Zeitschrift für Tierpsychologie*, 20, 410–33.

Tooby, J. and Cosmides, L. (1992). 'The psychological foundations of culture'. In J. Barkow, L. Cosmides and J. Tooby (eds.), *The adapted mind*, pp. 19–136. New York: Oxford University Press.

Trivers, R. (1974). 'Parent-infant conflict'. *Amer. Zool.*, 14, 249–64.

——— (1985). *Social Evolution*. Menlo Park CA: Benjamin/Cummins.

Turiel, E. (1998). 'The development of morality'. In W. Damon and N. Eisenberg (eds.), *Handbook of child psychology* (5th edn.), pp. 863–932. New York: Wiley.

Valentine, S. and Fleischman, G. (2004). 'Ethics training and businesspersons' perceptions of organizational ethics'. *J. Business Ethics*, 52, 381–90.

Walters, L. and Palmer, J.G. (1997). *The ethics of gene therapy*. New York: Oxford University Press.

Warnock, M. (1998). *An intelligent person's guide to ethics*. London: Duckworth.

Weatherall, D. (2006). *The use of non-human primates in research*. London: Academy of Medical Sciences.

Weeramantry, C.G. (2003) *Illegality of nuclear weapons*. Colombo: International Association on Lawyers against Nuclear Arms.

Williams, B. (1985). *Ethics and the limits of Philosophy*. London: Fontana.

Wittgenstein, L. (1923). *Tractatus Logico-Philosophicus*. London: Routledge & Kegan Paul.

Woodburn, J. (1982) 'Egalitarian societies'. *Man*, 17, 431–51.

Woodward, B. (2004). *Plan of attack*. New York: Simon & Schuster.

Wrangham, R.W. (in press). 'Why apes and humans kill'. In *Conflict: the 2005 Darwin College lecture series*. Cambridge: Cambridge University Press.

Wrangham, R.W. and Wilson, M.L. (2004). 'Collective violence: comparisons between youths and chimpanzees'. *Ann. N.Y. Acad. Sci.*, 1036, 1–24.

Wright, P.H. (1984). 'Self-referent motivation and the intrinsic quality of friendship'. *J. Social and Personal Relationships*, 1, 115–30.

Zwetsloot, G.I.J.M. and Marrewijk, M.N.A. (2004). 'From quality to sustainability'. *J. Business Ethics*, 55, 77–82.

INDEX